MAKESHIFT SHELTERS

Technical Notes on their Construction

The Nuclear Protection Advisory Group with the
permission of and in association with the Swiss
Federal Office for Civil Defence

THE OCTAGON PRESS
LONDON

Copyright The Octagon Press Ltd, 1983

No part of this publication may be
reproduced by any method at present known,
or to be invented or adapted, without
express permission in writing from
the Copyright Owners,
The Octagon Press Limited,
whose registered address is :
Lynton House,
7–12 Tavistock Square,
London WC1H 9LT
England

ISBN 0 86304 028 4

Printed and bound at
The Camelot Press Ltd, Southampton

CONTENTS

INTRODUCTION

The Nuclear Protection Advisory Group asked the Swiss Federal Authority's Office of Civil Defence to collaborate in the English publication of these instructions for the adaptation of existing premises to makeshift shelters (TA BSR 1977).

Rather than trying to adapt the Swiss version to English conditions, we are publishing a translation of the original text with some omissions and additions. Neither NuPAG nor the Swiss Federal Office of Civil Defence can be held responsible for any technical consequences of these instructions, but would advise readers to adjust the instructions to local conditions and to use them in this way as a guide. This publication is very much in line with the Group's aim to make information on matters of protection widely available.

NuPAG wishes to thank Mr Fritz Sager, Vice-Director of Swiss Civil Defence, for his assistance and co-operation in this project.

DEFINITIONS

Weapons effects	An all-embracing term covering the direct and indirect effects of nuclear weapons, chemical warfare agents and conventional weapons on the shelter and/or its occupants. (These weapons effects and general protective measures to counter them are discussed in some detail in Appendix 2.)
Mechanical effects	An all-embracing term covering the effects of blast, ground pressure, vibration, fragments, rubble, etc.
Fire risk	A quantitative indication of the movable and immovable combustible material in the immediate vicinity of the shelter.
Protective effect, protective scope	The totality of all weapons effects that a shelter is capable of withstanding to a certain degree.
Degree of protection	A measure of the protection afforded against a weapon effect with which the shelter still just remains serviceable.
Protected area	Area or areas which offer a certain amount of protection.
Shelter shell	The enclosing parts of the structure (ceilings, walls, floor, closures) that separate the protected area (shelter) from the unprotected surrounding or adjoining areas/rooms.

Enclosing walls	Walls of the shelter shell.
Escape	Escape from the shelter by the occupants without outside aid following an attack that produces rubble, debris, etc.
Emergency exit	A general term covering all kinds of exits used for escape when the surroundings have been destroyed.
Escape tunnel	An underground emergency exit leading away from the building.
Rubble zone	The ground area within which accumulations of rubble must be expected following destruction of the building above ground. On all sides of the building it extends to a distance of up to $\frac{1}{2}$ H from the wall of the building (H = eaves height of corresponding wall of the building).
Air vent, air opening	Special openings in the shelter shell for shelter ventilation.
Extent of adaptation of the makeshift shelter	The extent of adaptation (adaptation phase) defines the level of conversion of the makeshift shelter. The higher the level of conversion, the greater the protective effect of the makeshift shelter.

1. FUNDAMENTALS

1.1 *Purpose of Makeshift Shelters*

Until local authorities make full provision for civil defence, there will be a shortfall in shelters for the public. If a hostile act occurs before this shortfall is made up, many people can still be protected if existing underground facilities are selected for use as makeshift shelters in peacetime and then converted step by step and occupied during the period preceding the attack.

Protection afforded by makeshift shelters:
The objective in setting up makeshift shelters is to secure the maximum possible protection within a limited time using limited makeshift means. Since the starting point, i.e., the existing peacetime underground facility, will vary a great deal from one case to another (nature of the ceiling, walls, floor, existing openings, position in relation to the terrain, etc.) and since the available resources will also vary widely, there is no point in demanding any particular minimum protection standard for all makeshift shelters. The protection afforded by the makeshift shelter can be improved in stages by breaking down the conversion work on the basis of adaptation phases. Completion of adaptation phase 1 means considerable protection against radioactive fall-out and the heat produced by atomic explosions and against fragments; completion of adaptation phase 2 ensures considerable protection against falling rubble and a certain amount of protection against blast. The protection afforded by makeshift shelters is essentially less than that provided by proper purpose-built shelters.

1.2 *Selecting Suitable Premises*

Makeshift shelters are suitable existing underground rooms/areas such as cellars, garages, storerooms, etc., that are converted in an improvised manner and thus offer the occupants a relatively high degree of protection against the effects of weapons. The amount of materials and time involved in the conversion work can be minimized if suitable rooms/areas are chosen.

The following are the main characteristics of suitable premises:

— **underground location as far as possible;**
— **the protective shell is mainly of reinforced and ordinary concrete;**
— **the number of openings (entrances, exits, doors, windows, etc.) is as small as possible;**
— **the risk from the effects of fire and rubble from neighbouring buildings is as slight as possible – avoid premises in densely built old parts of cities (because the means of access and escape are at risk).**

Appendix 1 shows a system for fairly precise assessment of such premises. The more suitable the premises are according to this system, the greater the protection already offered and the smaller the amount of work needed to adapt the makeshift shelter.

Comment on selecting premises:
Often, the best way of producing a good makeshift shelter is to join forces with neighbours and select the best premises in the immediate vicinity and then prepare and use these together.

1.3 Preconditions

These instructions for the preparation of makeshift shelters assume that certain important preconditions are satisfied, namely that:
— planning and design have already been completed in peacetime by building experts (architect, structural engineer, draughtsman);
— the makeshift shelters are prepared by groups of workers under the supervision of building experts during the period preceding the attack.

1.4 Adaptation

The following chart indicates the preparatory work normally needed, arranged on the basis of adaptation phases. The protection afforded by the makeshift shelters increases the greater the level of adaptation.

Adaptation phase	Conversion work	Purpose
Adaptation phase 0	(1) Clear makeshift shelter (without carrying out any actual reinforcing)	Preparation of premises used in peacetime for occupation. Without reinforcement, a cleared makeshift shelter still offers some protection, particularly against radioactive fall-out and fragments.
Adaptation phase 1 Precondition: Completion of adaptation phase 0	(2) Seal up all openings in makeshift shelter, except entrance and emergency exit, using solid material. Install emergency lighting independent of mains power supply. (3) Reinforce entrance. Leave entrance opening large enough for bulky material (e.g., props, etc.) to be brought in easily during later conversion operations. (4) Seal up external openings in adjoining areas[1] with solid material. Close the doors, window shutters, roller shutters, blinds, etc., in the ground floor rooms above the makeshift shelter.	To provide better protection against radioactive fall-out and against the effects of fragments and fire. Emergency lighting in the event of a power failure.
	(5) Prepare emergency exit.	To provide means of escape in the event of fire or demolition of building above.
	(6) Produce controllable, i.e., closable air vents, Seal up remaining apertures (other openings).	To provide controllable natural ventilation. (Closure of air vents gives some protection against chemical warfare agents.)

Adaptation phase 2	(7) Support ceiling in makeshift shelter.	To increase load-bearing capacity to withstand rubble and pressure (blast) and any earth-filling.
	(7a) Reinforce ceilings of adjacent areas to some extent.	
Precondition: Completion of adaptation phase 1	(8) Reinforce exposed walls with earth embankments. (9) Reinforce entrances further as necessary.	To improve protection against radiation, fire and fragments.
	(10) Remove combustible material from areas directly above or near the makeshift shelter.	To reduce fire hazard (fire prevention precautions).
	(11) Extend emergency exit and ventilation ducting beyond the rubble zone (at least half eaves height).	To improve chances of escape and ventilation in the event of fire or demolition of the building above.
	(12) Reinforce ceiling of makeshift shelter (e.g., by covering with earth) or seal up external openings in ground floor areas above the makeshift shelter using solid material.	To improve protection against fire and radiation.
	(13) Install makeshift ventilation fan.	To improve air circulation in the makeshift shelter.

[1] Adjoining areas are areas or rooms directly adjacent to the makeshift shelter.

The internal fittings and equipment of makeshift shelters are described in section 4.

The maximum number of places in makeshift shelters is calculated on the basis of a volume of 4·0 cu.m for each occupant.

2. PLANNING

2.1 *Procedure*

The conversion work should be planned by building experts. This planning is vital for the detailed planning of the work and procurement of materials, and for the actual preparation of the makeshift shelter.

The detailed surveys mentioned in Appendix, 1.1 must be carried out first before starting on the planning. Having taken these surveys, re-check the suitability of the premises/structure for use as a makeshift shelter (with reference to Appendix 1). Then translate the reinforcing operations into:
— conversion plans (ground plan, sectional and detail plans)
— list of materials, and
— charts indicating times required (in a building schedule if desired).

Draw up a technical specification covering the information that is not clear from the plans and lists (e.g., building methods, possible variants, space requirement calculations, etc.).

The conversion operations should be set out in the order indicated in section 1.4. This means that the detailed information needed to complete the individual adaptation phases (material procurement, times required) is immediately available.

For planning purposes, sections 2.3 and 2.4 contain general information about material and time requirements. Part 3 provides notes on construction/conversion. Appendix 1 gives an example of a makeshift shelter project.

Important:

The risk of accident while setting up and using makeshift shelters should be precluded by careful planning as far as possible (secure fixtures to prevent them falling or being knocked over, eliminate the risk of tripping over objects, avoid projecting laths and protruding nails, etc.).

To avoid accidents while the conversion work is actually being carried out, observe the relevant regulations governing building and construction.

2.2 Basic Planning Requirements

Before planning the makeshift shelter, the following surveys must be made of the premises, i.e., the structure to be used as a makeshift shelter, the associated building, and the surrounding area. The data can also be obtained from the plans of the finished building project but the information must always be checked on the spot.

(1) Establish the ground plan of the areas to be used for the makeshift shelter including access passages and adjoining areas. Produce a scale ground plan if no construction drawings are available.

(2) Check the elevations (sections) of the areas to be used for the makeshift shelter. Record ceiling heights, location of openings in walls and the level of the ground as it affects external walls. Produce scale sectional plans if no construction drawings are available.

(3) Check the types of construction used in the enclosing parts of the structure (floors, walls, ceilings, doors, partition walls, etc.) including thickness and building materials used (reinforced concrete, unreinforced concrete, brick, metal, timber, etc.). Record on the ground plan and sectional drawings.

(4) Survey the building in which the makeshift shelter is located:

—Approx. height of the building (eaves height),
—Building type (modern construction, old construction with wooden floors, etc.).

When the makeshift shelter does not have any buildings above it (e.g., in the case of underground car parks under open spaces), the nature and thickness of the covering material (soil, paving, etc.) should be checked and recorded on the ground plan and sectional drawings.

(5) Survey the surrounding area (neighbouring buildings):

—Distance from other buildings,
—Heights of other buildings (eaves height),
—Types of buildings.

Record in extract from a plan of the locality or produce a sketch of the location.

6

(6) Note the fixed installations (ducting, piping, cables in building basement and makeshift shelter):
Record the most important information such as the nature of the medium being carried (water, waste water, gas, electricity), routing, main cocks, distribution points, etc., on the ground plan and sectional drawings. Fuel oil tanks in the building or in the surrounding area are not a threat to makeshift shelters and therefore need not be marked.

On the basis of these surveys, the suitability of the premises for use as a makeshift shelter and the number of places it offers must be checked again. (Compare with form in Appendix 1.)

2.3 Tools and Materials

Tools
The following tools above all are needed for the work:

— Tools for working with wood and metal such as axes, wood and metal saws, pincers/pliers, hammer, hand drill with accessories, etc.
— Tools for working with brick or stone such as hammers, mallets, chisels, etc.
— Shovels and picks, wheelbarrow.
— Electrical and lighting materials.

In addition, the following tools and equipment are useful:

— Percussion drill with accessories.
— Electrical equipment such as extension cables, plugs.
— Frame for filling sandbags (makeshift construction).
— Wheelbarrows, 'concrete buggies', cycle trailers and similar small conveyances.
— Small concrete mixer.
— Earth-moving equipment (e.g., conveyor, excavating machines, etc.).
— Welding equipment.
— Grader.

Materials

The main materials required for the conversion work are listed in the following chart:

Type of work	Useful materials
Propping and bracing, sealing up openings with solid materials, reinforcing doors	Round-section timber, angle-section timber, boards, shuttering panels, battens, hardwood wedges, steel ceiling-boarding supports or other steel supports, sandy gravel, cement (including rapid-setting cement possibly), old tyres for prop base supports, iron pipes (from 1″ to 3″ in diameter), small items such as wire, nails (80–140 mm in length), etc.
Sealing	Plastic sheeting, rubber bands, putty, battens, broad adhesive strip.
Reinforcing walls and ceilings with filling material or embankments	Earth, large strong sacks (to be filled with earth or sand), round-section timber, angle-section timber, boards (for use as shuttering).

2.4 *Time Required*

The time required to carry out the conversion work depends on the particular circumstances, the suitability of the building to be used as a makeshift shelter, the specialist knowledge or level of training of the building workers and on the extent to which adaptation is to be taken. The time needed to procure the materials is not included in the times indicated in the following tables.

Approximate time needed to carry out the conversion work

Suitability of premises (based on assessment in Appendix 1)	Maximum time needed in man-hours per sq.m of usable floor area of the makeshift shelter							
	For adaptation phase 0		For adaptation phase 1		For adaptation phase 2		Total time needed for adaptation phase 2	
	K[1]	G[2]	K	G	K	G	K	G
Very good	$\frac{1}{2}$	$\frac{1}{2}$	$1\frac{1}{2}$	1	2	$1\frac{1}{2}$	4	3
Good	$\frac{1}{2}$	$\frac{1}{2}$	$2\frac{1}{2}$	$1\frac{1}{2}$	5	3	8	5
Usable	$\frac{1}{2}$	$\frac{1}{2}$	$3\frac{1}{2}$	2	8	$6\frac{1}{2}$	12	9

[1] K means fairly small makeshift shelters with a floor area of up to approx. 50 sq.m
[2] G means larger makeshift shelters with a floor area exceeding about 50 sq.m

Detail times required

These times can be used as guide values when drawing up a chart for times required.

	Type of work	Unit	Approx. time needed in man-hours
Adaptation phase 0	Clearing premises used in peacetime, preparing for conversion and occupation (excl. fittings)	per building depending on size and purpose for which it is to be used	5–20
Adaptation phase 1	Sealing up openings with solid materials	per small opening less than 1 sq.m large openings of more than 1 sq.m (per sq.m)	2–4 3–5
	Reinforcing a door	per door	5–10
	Constructing an emergency exit with cover	per emergency exit	10–15
	Sealing gaps (around doors, etc.) and preparing sealable air vents	per shelter	3–5
Adaptation phase 2	Propping the ceiling including securing of props	per prop	1–2
	Prop base supports	per prop support	1–3
	Spreading earth over floor above ceiling of makeshift shelter	per sq.m floor area	1–3
	Reinforcing walls with loose earth embankments or 'walls' of sandbags	per sq.m of wall area	3–10 (depending on availability of plant)
	Constructing an emergency exit to a point beyond the rubble area	per emergency unit	20–50 (depending on availability of plant)
	Installation of makeshift ventilation fan	per fan	2–3

3. STRUCTURAL WORK

3.1 *Sealing Up Openings*

Openings in the enclosing walls (and ceilings) which are not being used for the entrance, emergency exit or air vent of the makeshift shelter must be sealed up with solid materials to ensure protection against the effects of weapons.

These openings should be sealed up in the course of adaptation phase 1. External openings in exposed areas adjoining the makeshift shelter must also be sealed up in adaptation phase 1.

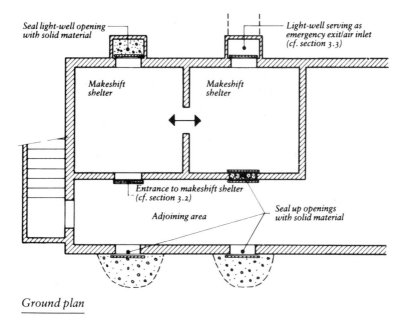

Ground plan

Fig. 1: *General view of openings to be sealed up with solid material*

If the floor above the makeshift shelter cannot be covered with earth, the external openings of the ground-floor rooms/areas above the makeshift shelter should be sealed up with solid material in the course of adaptation phase 2 (cf. also section 3.45 – Increasing protection against the effects of fire and nuclear radiation).

A few typical openings are shown in the following with the necessary strengthening.

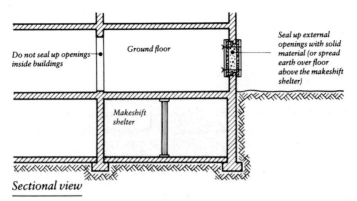

Sectional view

Fig. 2: Sealing up openings in ground-floor rooms with solid materials

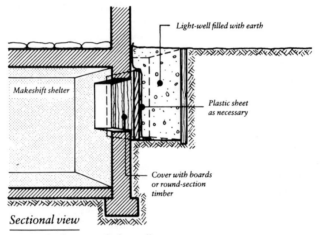

Sectional view

Fig. 3: Sealing up a light-well

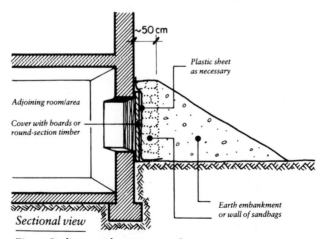

Sectional view

Fig. 4: Sealing up a basement window

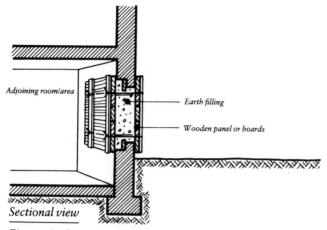

Adjoining room/area

Earth filling

Wooden panel or boards

Sectional view

Fig. 5a: Sealing up window openings with solid materials

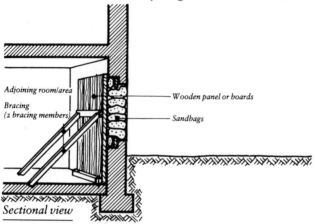

Adjoining room/area

Bracing
(2 bracing members)

Wooden panel or boards

Sandbags

Sectional view

Fig. 5b: Sealing up window openings with solid materials

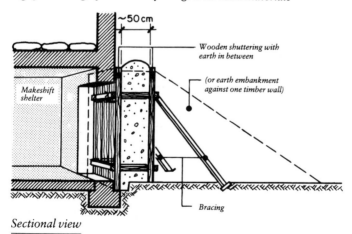

~50 cm

Wooden shuttering with
earth in between

(or earth embankment
against one timber wall)

Makeshift
shelter

Bracing

Sectional view

Fig. 6: Sealing up a large opening (e.g., garage door)

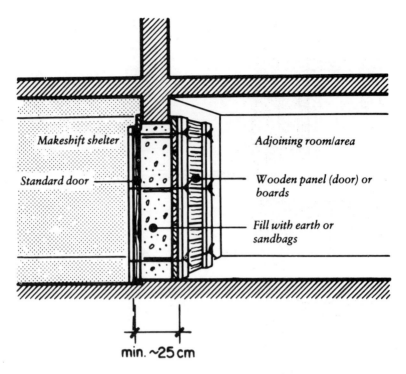

Makeshift shelter

Standard door

Adjoining room/area

Wooden panel (door) or
boards

Fill with earth or
sandbags

min. ~25 cm

Sectional view

Fig. 7: Sealing up a doorway with solid material

3.2 Reinforcing Entrances

As a rule the makeshift shelter should have only one entrance.
Superfluous peacetime doorways should be sealed up with
solid material as indicated in section 3.1. It is easier and more
effective to seal up openings with solid material than to
reinforce entrances with movable closures.

The remaining entrance to the makeshift shelter should be
reinforced as much as possible to protect against splinters, fire
and radiation. The entrance should be fitted with a reinforced
door as indicated in section 3.22 at least in the course of
adaptation phase 1.

3.21 Disposition of entrances

Where possible, only use 'internal' entrances so as to reduce
the amount of reinforcing work involved. These are entrances
leading to the makeshift shelter from covered adjoining
rooms/areas.

14

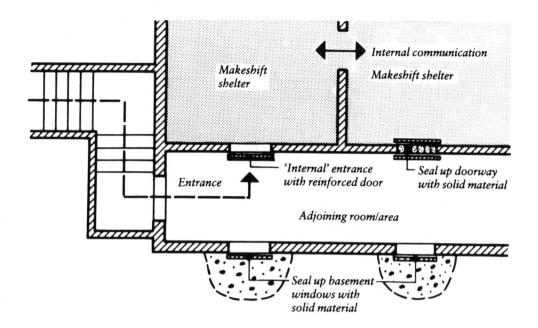

Internal communication

Makeshift shelter

Makeshift shelter

'Internal' entrance
with reinforced door

Seal up doorway
with solid material

Entrance

Adjoining room/area

Seal up basement
windows with
solid material

Ground plan

Fig. 8: Example of disposition of an 'internal' entrance

'External' entrances, i.e., entrances leading directly into the
makeshift shelter from the outside, should not be used unless
no 'internal' entrance is available. This is the case, for
example, with entrances into free-standing underground garages not having secondary entrances. As a rule, reinforcement
of external entrances involves a lot of work.

One way of reinforcing the entrance to a free-standing
underground garage is shown in the following example.

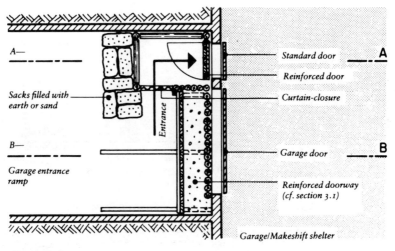

Standard door
Reinforced door
Curtain-closure
Garage door
Reinforced doorway
(cf. section 3.1)
Garage/Makeshift shelter

A—
Sacks filled with earth or sand
B—
Garage entrance ramp
Entrance

Ground plan

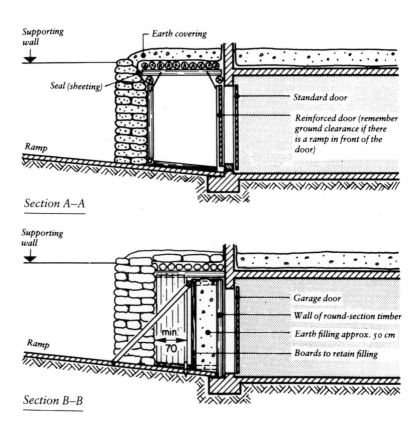

Supporting wall
Earth covering
Seal (sheeting)
Standard door
Reinforced door (remember ground clearance if there is a ramp in front of the door)
Ramp

Section A–A

Supporting wall
Garage door
Wall of round-section timber
Earth filling approx. 50 cm
Boards to retain filling
Ramp
min. 70

Section B–B

Fig. 9: *Example of an external entrance in combination with the garage entrance*

16

3.22 Reinforcing entrance doors

Entrance doors which open towards the outside and are made of steel or solid wood (at least 40 mm thick) for fire-retention in peacetime need not be reinforced.

Entrances with a standard door opening towards the inside or the outside must be reinforced. A few ways of strengthening such doors are illustrated in the following examples.

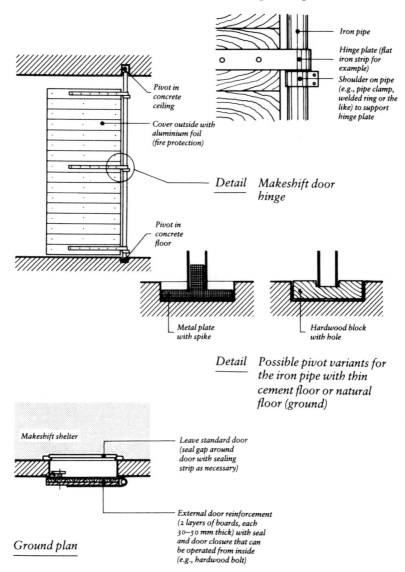

Iron pipe

Hinge plate (flat iron strip for example)

Shoulder on pipe (e.g., pipe clamp, welded ring or the like) to support hinge plate

Pivot in concrete ceiling

Cover outside with aluminium foil (fire protection)

Detail Makeshift door hinge

Pivot in concrete floor

Metal plate with spike

Hardwood block with hole

Detail Possible pivot variants for the iron pipe with thin cement floor or natural floor (ground)

Makeshift shelter

Leave standard door (seal gap around door with sealing strip as necessary)

External door reinforcement (2 layers of boards, each 30–50 mm thick) with seal and door closure that can be operated from inside (e.g., hardwood bolt)

Ground plan

Fig. 10: _Example of door reinforcement with standard door opening towards the inside_

With entrances having a standard door opening outwards, e.g., a door leaf made of wood less than 40 mm thick or of light timber construction or with recesses, this should be strengthened as in Fig. 11. When doing this, make sure first of all that the hinges can take the extra weight of about 70 kg added by virtue of the reinforcement by testing (hang from the door with your full weight!). In cases where boards cannot be nailed on or where the existing door hinges are not strong enough, remove the standard door with the door frame and fit a door reinforcement of the kind shown in Fig. 10.

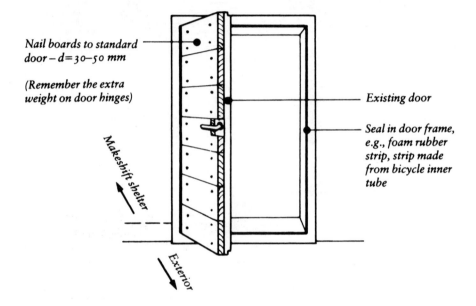

Nail boards to standard door – d=30–50 mm

(Remember the extra weight on door hinges)

Makeshift shelter

Exterior

Existing door

Seal in door frame, e.g., foam rubber strip, strip made from bicycle inner tube

Elevation

Fig. 11: *Example of door reinforcement for a standard door opening towards the outside*

3.3 Emergency Exits

As in standard purpose-built shelters, makeshift shelters should be provided with means of escape (emergency exits) other than the actual entrance. Such emergency exits should normally be combined with the controllable, sealable air vent. Makeshift shelters should have at least 1 emergency exit, shelters with room for more than 200 people at least 2.

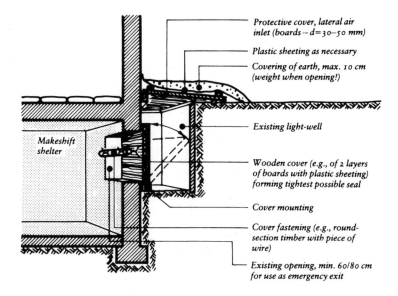

Protective cover, lateral air inlet (boards -- d=30-50 mm)

Plastic sheeting as necessary

Covering of earth, max. 10 cm (weight when opening!)

Existing light-well

Wooden cover (e.g., of 2 layers of boards with plastic sheeting) forming tightest possible seal

Cover mounting

Cover fastening (e.g., round-section timber with piece of wire)

Existing opening, min. 60/80 cm for use as emergency exit

Makeshift shelter

Sectional view

Fig. 12: Example of a combined emergency exit/light-well (air shaft) – (adaptation phase 1)

It must be possible to remove or push away the wooden cover and the protective cover from inside the makeshift shelter (escape).

Ways of extending the emergency exit to beyond the rubble zone (half the height h of the building above) are shown in the following figures (extend during adaptation phase 2).

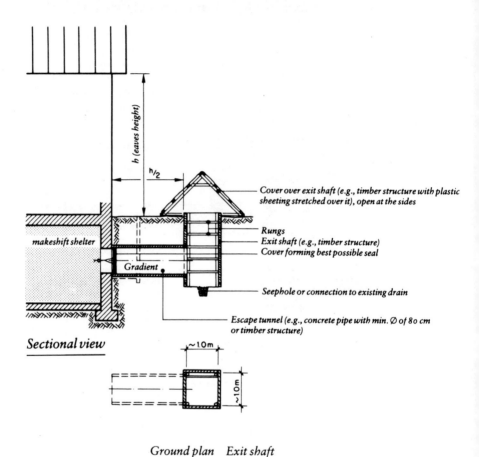

Sectional view

Cover over exit shaft (e.g., timber structure with plastic sheeting stretched over it), open at the sides

Rungs
Exit shaft (e.g., timber structure)
Cover forming best possible seal

Seephole or connection to existing drain

Escape tunnel (e.g., concrete pipe with min. Ø of 80 cm or timber structure)

Ground plan   Exit shaft

Fig. 13: Example of an emergency exit extending beyond the rubble zone h/2 (extend during adaptation phase 2)

20

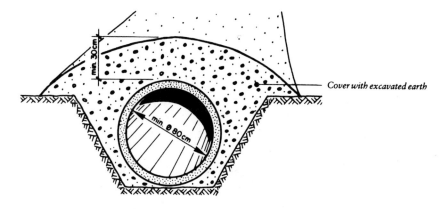

Cover with excavated earth

min. 30 cm

min. ∅ 80 cm

Sectional view

Fig. 14: Escape tunnel made of concrete pipe with minimum diameter of 80
cm or other strong tubular structure

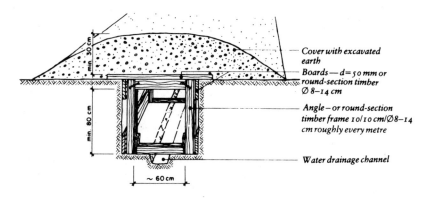

min. 30 cm

min. 80 cm

~ 60 cm

Cover with excavated
earth
Boards — d = 50 mm or
round-section timber
∅ 8–14 cm

Angle – or round-section
timber frame 10/10 cm/∅ 8–14
cm roughly every metre

Water drainage channel

Sectional view

Fig. 15: Timber escape tunnel (shored trench with covering material)

3.4 Reinforcing Ceilings

As a rule, ceilings of makeshift shelters are only reinforced in the course of adaptation phase 2.

The purpose of reinforcing ceilings:

— to increase the load-bearing capacity of the ceiling so that it can take the weight of the earth scattered over it and can withstand blast to a minor degree and the weight of the rubble produced by the latter.

Work involved: provision of additional ceiling support.

— to increase protection against:

the heat produced by fires above the makeshift shelter, nuclear radiation.

Work involved: increasing the mass (thickness) of the ceiling structure by covering it with earth or at least sealing up the external openings in the rooms/areas directly above the makeshift shelter.

Load-bearing capacity:

For this purpose, load-bearing capacity means failure load. For normal basement ceilings without additional support, this is about 1·0–1·5 t/sq.m overall (average 1·25 t/sq.m). This can be calculated roughly by taking the maximum peacetime loads (weight of the ceiling and maximum permissible use load) and multiplying the result by the safety factor of 2·0 to 2·5.

The load-bearing capacity of such ceiling slabs can be increased several times over by halving the spans by means of additional props or supports. As a result, a small amount of blast and the weight of the rubble produced by it can be withstood over and above the weight of the ceiling and any earth spread over it.

3.41 Increasing the load-supporting capacity of reinforced concrete ceilings of relatively small dimensions in building basements

(Ceilings supported on 2 to 4 sides, minimum ceiling thickness $d_{min} = 14$ cm)

The props are arranged and their dimensions are chosen as indicated in Figs. 16, 17 and the dimensional table Fig. 18.

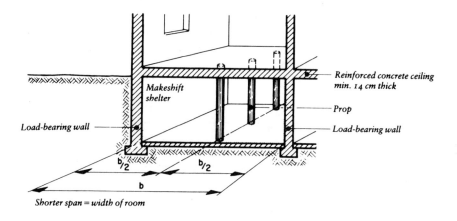

Reinforced concrete ceiling
min. 14 cm thick

Prop

Load-bearing wall

Makeshift
shelter

Load-bearing wall

$b/2$

$b/2$

b

Shorter span = width of room

Sectional view

Fig. 16: General view of prop arrangement

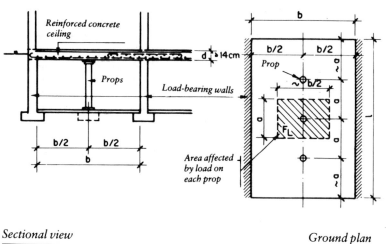

Reinforced concrete
ceiling

Props

Load-bearing walls

$b/2$

$b/2$

b

$d \approx 14\,cm$

b

$b/2$

$b/2$

Prop

$\sim b/2$

FL

Area affected
by load on
each prop

Sectional view

Ground plan

Fig. 17: Arrangement of props

Room width b (m)	Ø of round-section timber props (cm)	Rectangular timber props (cm)	Prop spacing a required (cm)
2·0	10	10 × 10	70
3·0	10	10 × 10	90
	12	10 × 12	90
4·0	10	10 × 10	45
	12	10 × 12	80
	14	12 × 14	100
5·0	10	10 × 10	40
	12	10 × 12	70
	14	12 × 14	110
6·0	10	10 × 10	—
	12	10 × 12	55
	14	12 × 14	90
	16	14 × 14	120
7.0	10	10 × 10	—
	12	10 × 12	45
	14	12 × 14	75
	16	14 × 14	110
	18	14 × 18	130
8·0	10	10 × 10	—
	12	10 × 12	40
	14	12 × 14	70
	16	14 × 14	100
	18	14 × 18	130
	20	16 × 20	140
9·0	10	10 × 10	—
	12	10 × 12	—
	14	12 × 14	60
	16	14 × 14	90
	18	14 × 18	120
	20	16 × 20	150

Fig. 18: Table of dimensions (for rooms with a height of 2·50 m or less)

The following must be noted with regard to these dimensional data:

— For ceilings where the ratio of the sides $b : l$ lies between $1 : 1$ and $1 : 1.5$, the additional props must be arranged – in the middle of the ceiling slab – in both ceiling support directions.
— Steel supports for ceiling boarding can also be used in place of the round timber props with diameters of 10 cm and 12 cm.
— The maximum prop spacing a indicated for specific widths of room must not be exceeded, even if round or angle-section props of greater thickness are used.

For rooms with a height exceeding 2·50 m, first select a prop spacing a with the aid of the table, Fig. 18, and then work out the approximate prop loading as follows:

$P \cong$ ($\sim$ ceiling failure load + weight of ceiling) $\times$ area over which load acts.

For this purpose, the ceiling failure load can be taken as approx. 5·0 t/sq.m (1·25 t/sq.m $\times$ 4 because span is halved) and the weight of the ceiling itself as approx. 0·7 t/sq.m

With this prop loading, the required diameter of round-section timber prop can be determined using the graph, Fig. 23. The extra props must be arranged as indicated in Fig. 17.

Examples of ceiling reinforcement calculation (room height 2·50 m or less):

(1) Given:
— room width b = 4·0 m
— room length l = 7·0 m
— room height h = 2·4 m
— ceiling is made of reinforced concrete, thickness d = 18 cm, i.e., more than d_{min} = 14 cm

Needed: — prop spacing a for a given prop diameter

Procedure: $b : l = 4·0 : 7·0 = 1 : 1·75 < 1 : 1·5 \rightarrow$ reinforcement is only needed in one direction. For a room of width b = 4·0 m and a prop diameter of 14 cm, the dimensional table, Fig. 18, gives a prop spacing of a = 100 cm.

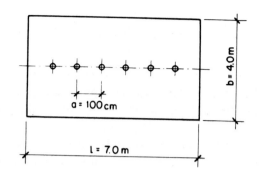

Ground plan

Fig. 19: Ground plan of prop arrangement

(2) Given: — room width $b = 5 \cdot 0$ m
            — room length $l = 5 \cdot 0$ m
            — room height $h = 2 \cdot 5$ m
            — ceiling is made of reinforced concrete, thickness $d = 18$ cm, i.e., more than $d_{min} = 14$ cm

Needed: — prop spacing a for a given prop diameter

Procedure: b : $l = 5 \cdot 0$: $5 \cdot 0 = 1 > 1$: $1 \cdot 5 \rightarrow$ reinforcement needed in both directions. For a room of width $b = 5 \cdot 0$ m and a prop diameter of 14 cm the dimensional table, Fig. 18, gives a prop spacing of $a = 110$ cm ($a = 100$ cm chosen).

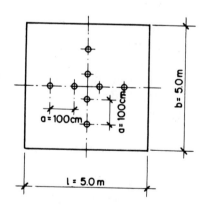

Ground plan

Fig. 20: Ground plan of prop arrangement

26

3.42　*Increasing the load-supporting capacity of flat-plate ceilings and beamed ceilings of reinforced concrete*

(Mimimum ceiling thickness $d_{min} = 14$ cm)

Ceilings carried on separate supporting columns made of concrete or steel (flat plate ceilings or beamed ceilings) are mainly to be found in underground garages and storerooms. Such ceilings often have a relatively high failure load – because of the the high use loads they are designed to take (office buildings) or the existing covering of earth. For this reason, ceilings with a permissible use load of more than 1500 kg/sq.m in peacetime and ceilings with a peacetime earth covering of more than 1 m thick need not be reinforced with additional props.

All other flat-plate and beamed ceilings should be reinforced by installing extra props in the course of adaptation phase 2. Dimensional information and arrangements for extra props for flat-plate and beamed ceilings are provided below. For this purpose, it is presumed that the existing supporting structure has been built to the required (British) standards. We cannot go into special structures here. However, in specific cases, the load-bearing characteristics and required reinforcements for a special ceiling can be ascertained from the designer's calculations or the building plans (reinforcement).

Flat-plate ceilings

Reinforced concrete ceiling on separate supporting columns (with or without a flared head)

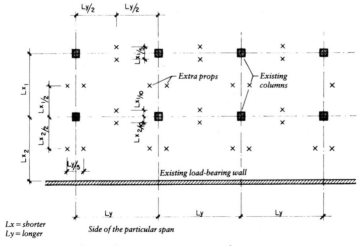

Lx = shorter
Ly = longer　　*Side of the particular span*

Fig. 21: *Ground plan showing arrangement of extra props*

The extra props must be arranged as indicated in Fig. 21. The prop dimensions table, Fig. 22, gives the required diameters of round-section timber props for reinforcing a flat-plate ceiling with a room height of 2·5 m or less.

Lx (m)	Ly (m)	Min. ceiling thickness (cm)	Ø of round-section timber prop (cm)
4·0	4·0 6·0	14 16	14 16
5·0	5·0 7·5	16 18	16 18
6·0	6·0 9·0	18 22	18 22
7·0	7·0 10·5	22 26	22 24
8·0	8·0 12·0	26 30	24 26
9·0	9·0 13·5	32 34	26 28

Fig. 22: Dimensional table (for room heights of 2·50 m or less)

For room heights in excess of 2·50 m, first work out the approximate load on the extra props as follows:

$$P \cong (\sim \text{ceiling failure load} + \text{ceiling weight}) \times Lx \times Ly \times 1/8$$

For this purpose, the ceiling failure load can be taken as $\sim 5\cdot0$ t/sq.m ($1\cdot25$ t/sq.m $\times 4$ because the span is halved) and the actual weight of the ceiling as $\sim 0\cdot7$ t/sq.m.

With this prop loading, the required diameter of round-section timber prop can be ascertained from the graph in Fig. 23. The extra props must be arranged as indicated in Fig. 21.

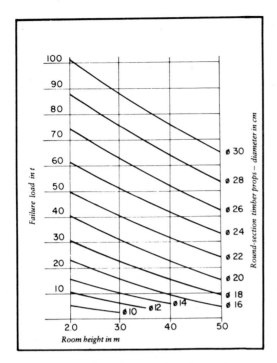

Fig. 23: *Failure load for round-section timber props*

Example for working out the reinforcement for a flat-plate ceiling:

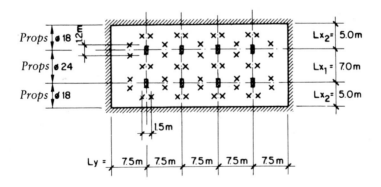

Fig. 24: *Ceiling dimensions and ground plan of prop arrangement*

Given: — room dimensions 37.5×17.0 m, with 8 existing separate supporting columns
              — room height approx. 2.4 m
              — ceiling thickness $d = 25$ cm (reinforced concrete)
              use load $p = 500$ kg/sq.m

Needed: — cross-sections of the extra props and their arrangement

Procedure: — For the prop cross-sections, the prop dimensions table, Fig. 22, indicates:

$Lx_1 = 7.0$ m and $Ly = 7.5$ m   Round timber prop $\emptyset$ 24 cm
$Lx_2 = 5.0$ m and $Ly = 7.5$ m   Round timber prop $\emptyset$ 18 cm

        — The arrangement of the extra props indicated by Fig. 21 is as follows:

$1/5 \times Ly = 1/5 \times 7.5$ m $= 1.5$ m
$(1/10 \times Lx_1) + (1/10 \times Lx_2) = 0.7 + 0.5 = 1.2$ m

Beamed ceilings

For systems with beams, the approximate load on the extra props under the beam is determined as follows:

$P \cong$ ($\sim$ ceiling failure load + ceiling weight) $\times$ area affected by load at extra beam props

For this purpose, the failure load and the weight of the ceiling itself can be taken to be the same as in the previous section $(5.0 + 0.7 = 5.7$ t/sq.m). With this prop loading, the required diameter of timber prop can be established from the graph, Fig. 23.

The arrangement and the size of the extra props under the ceiling slab must be as indicated in Fig. 25 and the table of dimensions, Fig. 18.

System with beams alone

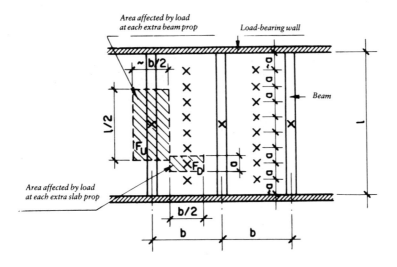

Area affected by load at each extra beam prop

Load-bearing wall

Beam

Area affected by load at each extra slab prop

Fig. 25: Ground plan showing arrangement of extra props

Example:

Given:
— Beam spacing $b = 5 \cdot 00$ m
— Beam span $l = 10 \cdot 00$ m
— Ceiling thickness 24 cm, i.e., more than $d_{min} = 14$ cm
— Room height $2 \cdot 5$ m

Needed:
— Prop spacing a
— Diameters of timber props under beam and slab

Procedure: — Prop under beam

$$P = \frac{b}{2} \cdot \frac{l}{2} \cdot (5 \cdot 0 + 0 \cdot 7) = \frac{5 \cdot 0}{2} \cdot \frac{10 \cdot 0}{2} \cdot 5 \cdot 7 = 71 \cdot 25 \text{ t}$$

The graph, Fig. 23, indicates a prop diameter of something over 26 cm (26 cm chosen).

— Prop under slab
$b : 10 \cdot 0 = 5 \cdot 0 : 10 \cdot 0 = 1 : 2 \rightarrow$ reinforcement only needed in one direction. For a room of width $b = 5 \cdot 0$ m and a prop diameter of 14 cm, the dimensional table, Fig. 18, indicates a prop spacing of $a = 110$ cm.

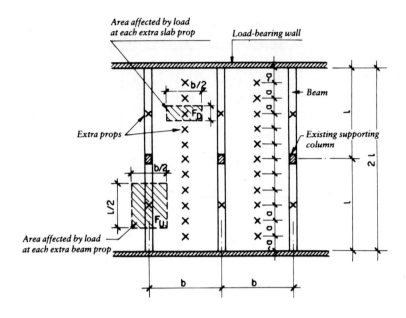

Fig. 26: Ground plan showing arrangement of extra props

Example:

Given: — Beam spacing $b = 6\cdot00$ m
— Beam span $l = 7\cdot50$ m
— Ceiling thickness 26 cm, i.e., more than $d_{min} = 14$ cm
— Room height $3\cdot0$ m

Needed: — Diameter of timber props under beam and slab
— Prop spacing a

Procedure: — Prop under beam

$$P = \frac{b}{2} \cdot \frac{l}{2} \cdot (5\cdot0 + 0\cdot7) = \frac{6\cdot0}{2} \cdot \frac{7\cdot5}{2} \cdot 5\cdot7 = 64 \text{ t}$$

The graph, Fig. 23, indicates a prop diameter of 26 cm.

— Prop under slab

$b : 1 = 6{\cdot}0 : 15{\cdot}0 = 1 : 2{\cdot}5 \rightarrow$ reinforcement only needed in one direction. For a room of width $b = 6{\cdot}0$ m and a prop diameter of 16 cm, the dimensional table, Fig. 18, indicates a prop spacing of $a = 120$ cm. Since the height of the room is $3{\cdot}00$ m the prop diameter must be checked:

Area affected by load

$$F_D = \frac{b}{2} \cdot a = 3{\cdot}0 \cdot 1{\cdot}2 = 3{\cdot}6 \text{ sq.m}$$

Existing prop loading: $3{\cdot}6 \cdot 5{\cdot}7 = 20{\cdot}5$ t.

The graph, Fig. 23, indicates a prop diameter of 18 cm.

3.43 Structural details for prop installation, prop base supports

The round or angle-section props are cut to the height between floor and ceiling less an amount for the prop capping and base support and installed and wedged in the predetermined position. The props must be secured so that they cannot topple over (e.g., as a result of the wedges coming loose or due to shaking). This can be achieved by locating retainers directly in the ceiling, securing the props to the ceiling, or by cementing them into a special prop base support.

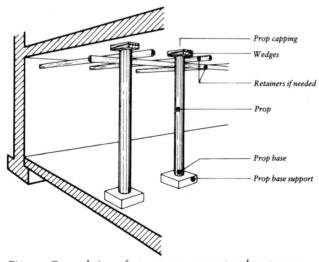

Fig. 27: General view of a prop arrangement and part names

Various kinds of prop capping are depicted in Figs. 28a to 28c.

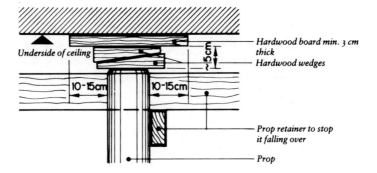

Fig. 28a: Prop capping with wedges and retainers

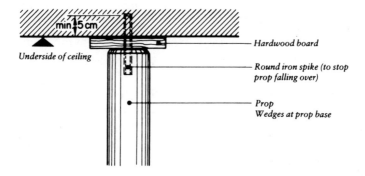

Fig. 28b: Prop capping with round iron spike

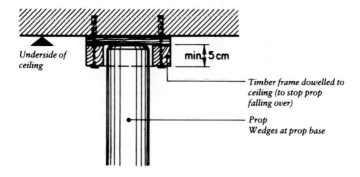

Fig. 28c: Prop capping with timber frame

Figures 29a and 29b show base supports standing on an existing concrete floor slab with a minimum thickness of 15 cm.

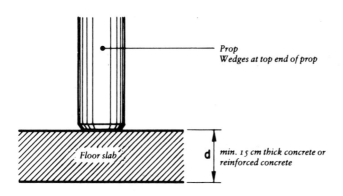

Fig. 29a: Prop base without any wedges

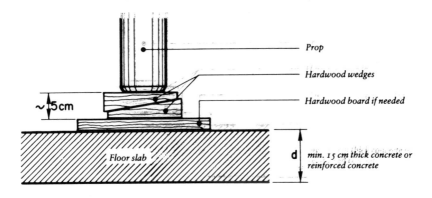

Fig. 29b: Prop base with wedges

With concrete floors less than 15 cm thick or natural earth floor, the props must be placed on specially prepared supports. Solutions that can be used in such cases are illustrated in Figs. 30a to 30c.

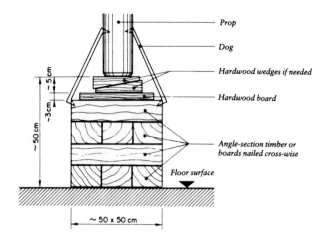

Fig. 30a: Prop base with wedges on angle-section timber base support

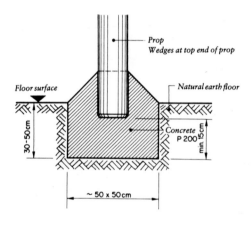

Fig. 30b: Prop base support made of concrete let into the floor

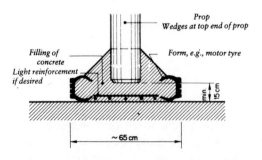

Fig. 30c: Prop base support concreted to the floor

36

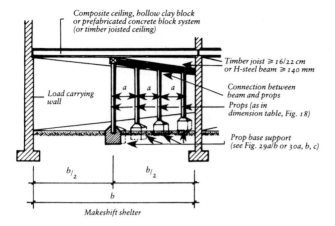

Composite ceiling, hollow clay block
or prefabricated concrete block system
(or timber joisted ceiling)

Timber joist ≥ 16/22 cm
or H-steel beam ≥ 140 mm

Connection between
beam and props

Props (as in
dimension table, Fig. 18)

Prop base support
(see Fig. 29a/b or 30a, b, c)

Load carrying
wall

a a a

$b/_2$ $b/_2$

b

Makeshift shelter

Fig. 30d: Timber calculation

3.44 *Remarks on increasing the load-supporting capacity of special ceilings* (e.g., composite ceilings, vaulted ceilings, etc.)

A statistical analysis must always be made when reinforcing such ceilings.

— Composite ceilings, hollow clay block or prefabricated concrete block systems can be strengthened following the principle of halving the spans in the main bearing axis by means of beams and props. (Props as in the dimensional table, Fig. 18.)
— Vaulted ceilings (usually made of brick) need not be strengthened as long as they are in good condition and stand on solid abutments. Do not prop arches.

3.45 *Increasing protection against the effects of fire and nuclear radiation*

— Ceilings without a building above and ceilings under buildings constituting a slight fire hazard [1] must be covered with a layer of earth or sandbags so as to give a total thickness of 30 cm.
— Ceilings under buildings constituting a major fire hazard [2] or in buildings where a relatively high proportion of the ground floor walls is taken up by openings (more than about 50% of the external wall area constituting openings) must be covered with a layer of earth or sandbags to give a total thickness of 40 cm.   [*Footnotes overleaf.*

37

— If cogent reasons make it impossible to cover the floor above the makeshift shelter with a layer of earth or sandbags, the external openings in rooms or areas directly above the shelter should be sealed up with solid material (cf. section 3.1).
— Fixtures and materials constituting a fire risk must be removed from adjoining areas above and on the same level as the makeshift shelter.

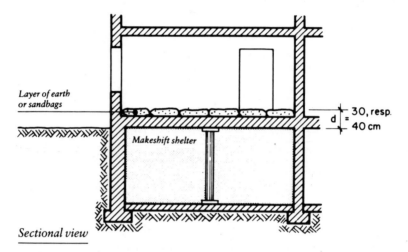

Fig. 31: *Reinforcing (thickening) the ceiling/floor with a layer of earth or sandbags*

3.46 *Reinforcing the ceilings of rooms/areas adjoining the makeshift shelter*

In the course of adaptation phase 2, the ceilings of the adjoining rooms/areas must be strengthened with props to withstand loadings produced by rubble to a distance of at least

[1] The following are 'slight fire hazards'
  1. One-family houses, excl. timber houses (floors and walls in wood).
  2. Blocks of flats and business premises of fireproof construction (except in 'Major fire hazard', 3).
[2] The following are 'major fire hazards': Quantities of combustible materials near shelter, such as
  1. more than one wooden floor/ceiling directly above the shelter,
  2. surface structure made of timber,
  3. combustible materials such as for example timber, furniture, plastics, hay, fuels, stored in one of the two storeys directly above the shelter or in areas near the shelter.

$1\cdot5 \times h$ (h = room height) from the wall of the makeshift shelter (procedure as in sections 3·41 to 3·44). This will prevent these areas being filled with rubble if the building collapses and thus ensure that the walls of the makeshift shelter are not subjected to horizontal pressure by rubble (ground pressure arising from rubble). It is not necessary to increase the protection afforded by the ceilings of neighbouring rooms against the effects of fire and nuclear radiation.

The ceilings of adjoining rooms/areas need not be reinforced where the wall of the makeshift shelter is made of concrete or reinforced concrete.

3.5   Reinforcing Walls

The enclosing walls of the makeshift shelter must afford some protection against:

— rubble and splinters/fragments
— the effects of fire and
— nuclear radiation.

Strengthening to withstand direct blast loadings would be very costly and, therefore, is not dealt with in these instructions.

Various typical wall arrangements are illustrated in the following and the normal reinforcing measures are described.

3.51   Outer walls below ground

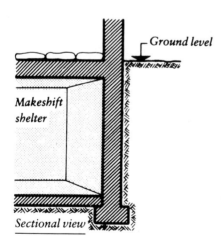

Fig. 32: Entire wall below ground – no wall reinforcement needed

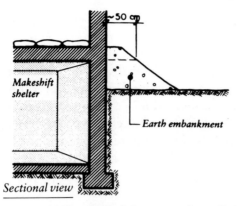

Fig. 33: Wall partially below ground – wall should be reinforced with an earth embankment or possibly a 'wall' of sandbags

3.52 Free-standing outer walls

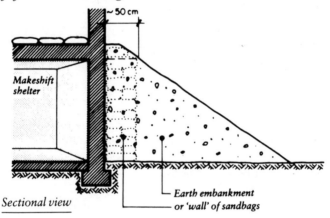

Fig. 34a: Wall reinforced with earth embankment or sandbags. When outer wall is of brick, wall must only be reinforced with sandbags (to avoid the extra load placed on the wall by an earth embankment).

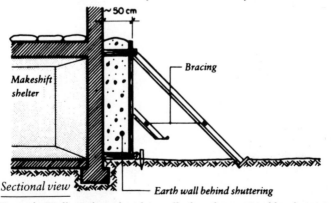

Fig. 34b: Wall reinforced with a wall of earth contained by shuttering

3.53   Free-standing interior walls

For free-standing interior walls, one of the reinforcing proce-
dures shown in Figs. 35 and 36 must be adopted:

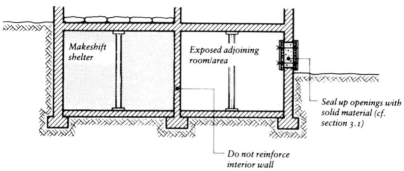

Sectional view

Fig. 35: *Seal up openings in neighbouring outer walls with solid material*

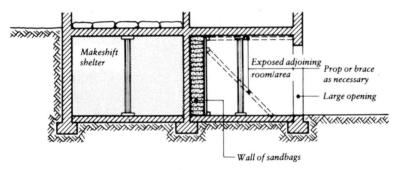

Sectional view

Fig. 36: *Reinforced interior wall. This method is advisable when the
neighbouring outer walls have very large openings which would
prove very costly to seal up with solid material*

3.54   Partition walls inside the shelter

Partition walls inside the actual makeshift shelter should
normally be left as they are.

Openings – e.g., for internal communication between rooms
within the shelter – should not be made unless the number of
entrances to the makeshift shelter can be reduced as a result.
(Whenever possible, provide just one shelter entrance, cf.
section 3.2.)

3.6   *Ventilation*

Shelters accommodating people for any length of time (hours, days) must be ventilated to ensure an adequate supply of fresh air for breathing.

A special ventilation system is provided in purpose-built shelters located in the underground basements of new buildings in Switzerland to provide the occupants of the shelter with a supply of fresh air. With this ventilation system the temperature and the humidity, in addition to the quality of the air, can be kept within acceptable limits for a protracted stay. This ventilation system for purpose-built shelters is also fitted with a gas filter to prevent the entry of any chemical warfare agents. As a rule, the incoming stream of air is only passed through this gas filter (active carbon filter) as a precaution when there is a danger that chemical warfare agents may be used. Collective protection against gas is obtained when the gas filter is used with the shelter closed and sealed.

Radioactive fall-out essentially occurs in the form of sandlike particles which give off radioactive radiation. These relatively heavy fall-out particles are not drawn into the shelter by the ventilation system, they settle in front of the air vent or, at the latest, in the air duct outside the shelter and thus do not pose any significant danger to the occupants of the shelter through the air supply system. Consequently, there is no need to use the collective gas filter in purpose-built shelters in the event of radioactive fall-out.

It is not advisable to install special ventilation systems in makeshift shelters for the following reasons:

There is no way of being sure how such systems will work in a makeshift shelter as far as protection against gas is concerned since the selected makeshift premises will not have a completely sealed shelter shell. Moreover, the cost of installing a special ventilation system is much too high in relation to the benefit that could be obtained.

The following is suggested for ventilating the makeshift shelter:

To make sure that a lengthy stay is possible, even in a makeshift shelter, the makeshift shelter is occupied to a lower density (4 cu.m per person instead of the 2·5 cu.m per occupant applicable to purpose-built shelters). In addition, the makeshift shelter should have natural ventilation with maximised control. To this end, steps should be taken to ensure that the air can only enter and leave through specific openings. The air vents should be constructed in such a way that they can be closed so that the flow of air through the makeshift shelter can be stopped completely. In addition, all the remaining openings and gaps must be closed off as tightly as possible. Putty, plasticine, plastic sheet, rubber, foam strip, adhesive strip, loam, etc., can be used as sealing materials. Visual inspection will give some indication of the quality of the sealing. The makeshift shelter will be largely sealed if no daylight can be seen when the lighting is turned off.

In a makeshift shelter, protection against gas must be obtained by the occupants taking precautions individually, i.e., by putting on a mask. The ventilation measures do not put makeshift shelters significantly at risk in the event of radio-active fall-out.

The following types of ventilation can be used without gas protection to ensure the air supply is adequate for breathing in makeshift shelters:

— Natural ventilation – which will normally be the case.
  In this case ventilation is effected by the natural draught between the air vents.

— Natural ventilation assisted by a makeshift fan installation. The installation of a makeshift fan[1] is a great advantage. Depending on the nature and position of the fan, the makeshift shelter can be forcibly supplied with fresh air or cleared of stale air, i.e., fresh air from outside can be sucked in or the stale air in the shelter can be blown out. As a result, controlled renewal of the air inside the makeshift shelter is assured. (This assumes the power supply is intact; natural ventilation occurs in the absence of power.)

[1] E.g., kitchen fan of type fitted in window or wall. (Kitchen fans vary in throughput from 300 to 400 cu.m/h.) Desirable throughput:
  at least 100 cu.m/h for makeshift shelter with up to 25 places
  at least 250 cu.m/h for makeshift shelter with up to 50 places
  at least 500 cu.m/h for makeshift shelter with up to 100 places.

— Natural ventilation and use of existing ventilation system. When makeshift shelters are set up in underground garages, it is sometimes possible to use the existing peacetime garage ventilation system. In such cases the foul-air and fresh-air vents should be fitted with covers that can be closed quickly in the event of the system stopping.

Figures 37 to 39 show general schemes for natural ventilation and for natural ventilation assisted by a makeshift fan installation.

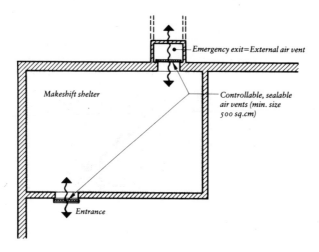

Fig. 37: Natural ventilation

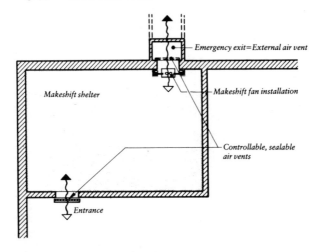

Fig. 38: Natural ventilation assisted by makeshift fan installation

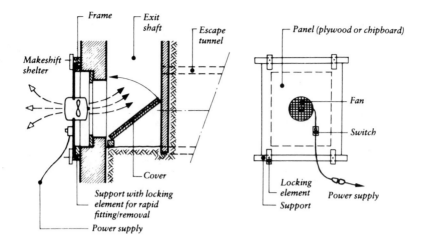

Sectional view   *View from inside*

Fig. 39: Fan installation

4. INTERNAL FITTINGS AND EQUIPMENT

4.1 Reserve Supplies of Water

A reserve of at least 10 to 20 litres of water per person should be laid in for drinking and essential washing purposes in case of a protracted stay (a few days) in the shelter (physiological minimum requirement for survival is approx. 1 to 2 litres per person per day assuming a diet of dry food).

Drinking water should be stored in containers that do not pose any health risk, e.g.:

Container	Storage
Water canisters	On the ground, stacked
Camping containers, water bags	On the ground or suspended
Buckets, pails, other suitable household containers	On the ground or in racks near the ground
Containers with discharge taps, metal or plastic-lined receptacles	On the ground on low stands
Bottles	In crates or boxes on the ground

Suitable ladles or containers should be provided for using the water. The water should be stirred from time to time. Proprietary water sterilization tablets should be used to keep the water pure.

4.2 Makeshift toilet

Every shelter should have a makeshift toilet (dry toilet) in case of a protracted stay – one toilet at least for every 30 people. Proprietary toilets, including versions for camping, or makeshift toilets can be used for the dry toilet.

Instructions on how to make an improvised dry toilet:

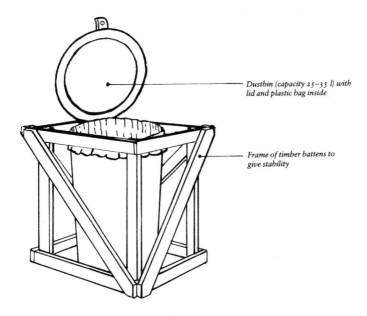

Dustbin (capacity 25–35 l) with lid and plastic bag inside

Frame of timber battens to give stability

Fig. 40: Makeshift toilet

In addition, cleaning materials, newspaper, earth or chloride of lime (bleaching powder) or similar should be provided to cover the contents. The toilet can easily be partitioned off by hanging a curtain from the ceiling. If possible, the makeshift toilet should be placed near the stale-air outlet vent.

4.3   Lighting

Normally the *existing lighting* in the shelter should be used (supplemented by additional lamps if necessary).

Emergency lighting: A minimum level of lighting must be ensured at all times, even in the event of a power failure. The emergency lighting can be improvised with the following means by way of example:

— battery-powered lamps – car batteries can also be used;
— candles and similar kinds of lamp (CAUTION: Do not use too many candles, gas lamps, oil lamps, etc., because they consume oxygen. The condition of the air should be monitored continually with the aid of a burning candle. One candle consumes roughly the same amount of oxygen as a man).

4.4 Provision for Sleeping, Waking Hours, Effects

A minimum of about 1·5 to 2 sq.m of floor area per person is provided in makeshift shelters. This is usually enough for lying or sitting on the floor (e.g., on mattresses, air-beds, blankets and the like) and to accommodate personal effects, stocks of water and rations etc.

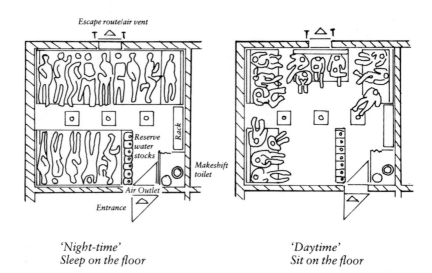

'Night-time'
Sleep on the floor

'Daytime'
Sit on the floor

Fig. 41: Disposition of occupants

4.5 Further Items

Source of information:
Transistor radio (remember spare batteries!) with wire aerial run outside.

Implements to facilitate escape:
Axe, saw, pick-axe, spade, crowbar, etc.

Fire-fighting equipment (near shelter):
Sand, water, broom with wet jute sacking, wet towels.

First-aid kit:
First-aid box, e.g., first-aid kit from a car, or similar.

Pack and rations:
Every person will need the following personal effects and rations for a stay of a few days:

Pack:
The following should be packed in trunks, rucksacks, cases or other suitable containers bearing the owner's name in case it is necessary to stay in the shelter for a few days:
Underwear, socks, handkerchiefs, track/training suit or trousers and pullover, plimsolls or slippers, sleeping bag or blanket, foam-rubber mattress or air-bed, toilet bag, hand towel, flannel, WC paper, plastic bags (25–35 l capacity, at least 10 in number), personal medicines, sewing materials, cord, torch with spare batteries, thick candles, matches, canteen, penknife, cutlery, plate and cup (plastic), writing implement, note-paper, entertainment (books, games), personal documents, money and valuables (when not deposited in a bank), ration cards (if required).

Personal rations:
To be eaten cold in the shelter. Keep ready, properly packed in a case or box. *For about 1 week*, a person needs an intake of around 50,000 to 60,000 kJ.

Example:	Total per person
Sugar, chocolate, nuts	1 kg
Preserved meat/fish, cooked sausage	1 kg
Preserved vegetables/fruit	1 kg
Crisp-bread, biscuits, confectionery	$\frac{1}{2}$ kg
Salted butter, margarine, processed cheese	$\frac{1}{2}$ kg
Jams, spreads, dried fruit	$\frac{1}{2}$ kg
Sweetened condensed milk	$\frac{1}{2}$ kg
Syrup/treacle	$\frac{1}{2}$ kg

Salt, pepper, instant coffee, instant tea (tea-bags?), chewing gum, sweets, vitamin preparations, etc., as required.

5. EXAMPLE OF A MAKESHIFT SHELTER

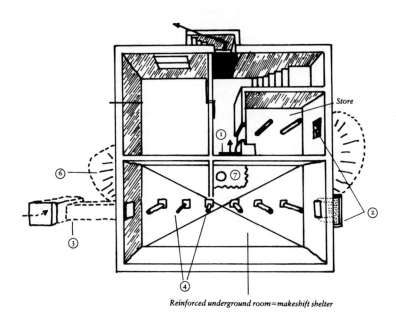

Reinforced underground room=makeshift shelter

Fig. 42: View from above into the basement of a single-family (detached) dwelling

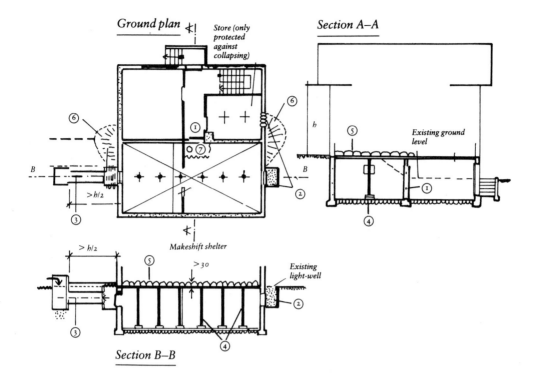

Ground plan

Store (only protected against collapsing)

Section A–A

Existing ground level

Makeshift shelter

Section B–B

Existing light-well

Key for Figs. 42 and 43:
(1) Reinforced entrance door and air vent
(2) Openings sealed up with solid material
(3) Escape route and air vent
(4) Ceiling supported with props on strengthened base
(5) Ceiling reinforced (thickened) with earth or sand
(6) Enclosing walls supported with earth embankments
(7) Makeshift toilet

Fig. 43: Ground plan and sections

51

6. NOTES ON THE USE OF MAKESHIFT SHELTERS

If the makeshift shelter is to be used correctly and full advantage is to be taken of the protection it offers, people seeking shelter must keep as calm as possible in assessing the situation and thus the danger indicated by warnings issued by the authorities, bulletins in the mass media (radio, television, newspapers) or personal perception. It is better to seek the protection of the makeshift shelter more often and stay there longer rather than take chances. Protracted stays in shelters can be broken up by brief periods outside depending on the danger outside at a particular time (rotation). Particular care must be taken with regard to weapons effects which the individual may hardly notice, if at all, until it is too late. Therefore, after a warning has been given, nobody should leave the makeshift shelter until reports confirm that no chemical warfare agents have been used within a fairly wide radius and that there is no radioactive fall-out. Even after an attack the shelter is the safest place to be – even when the building has collapsed.

A few significant situations are illustrated in the following and the appropriate steps are briefly outlined.

Situation	Measures inside the makeshift shelter

Preparation of the makeshift shelter
In times of great political and military tension (danger of war)
	— Prepare the makeshift shelter as indicated in sections 1 and 2 of these instructions.
	— Lay in stocks of water and rations for a stay of a few days.
	— Convert according to available time, materials and labour.
	— Take steps to prevent fire:
	  — Remove combustible materials from rooms or areas immediately above or adjacent to the makeshift shelter.
	  — Close the windows, shutters and blinds in the building to make it harder for fire to break out.
	  — Prepare fire-fighting and extinguishing materials.

Occupation of the makeshift shelter As a precaution, move into the shelter as soon as hostilities begin because with modern weapon systems ample warning may often be impossible	— Move into the makeshift shelter. — Check shelter ventilation as indicated in section 3.6. — Listen to the radio. — Constantly improve reinforcement and level of equipment of the makeshift shelter as outside danger allows. — Top up stocks of water, stir as necessary.

Protracted stay in the makeshift shelter
This may be necessary:
— in the event of protracted severe threat from possible weapons effects,
— following an attack when the after-effects last for some time, e.g., radioactive fall-out
	— Establish and follow a daily routine, arranging:
	  — catering
	  — resting times
	  — occupations
	  — work in the makeshift shelter such as cleaning, including toilets
	  — replenishment of stocks (of water, food) where possible
	  — watch duties: listening to the radio, checking the quality of the air by watching the burning candle
	  — rotation, i.e., short periods outside the makeshift shelter (depending on danger outside)
	  — care and medical treatment for those needing attention.
	
	Note on catering inside the makeshift shelter: As a rule, cooking should not be done in the shelter itself because of the temperature and humidity and the quality of the air for breathing. Depending on the danger outside the shelter, conveniently located rooms or even nearby kitchens can be used for cooking. Otherwise, occupants must put up with cold food temporarily.

Chemical attack warning or surprise chemical attack i.e., immediate use of chemical warfare agents is expected or chemical warfare agents have already been used in the vicinity (local alert, information over	— Put on protective mask (gas mask) if available. — Do NOT leave makeshift shelter. — Close all vents and openings in the makeshift shelter at once and seal them. Stop any artificial ventilation immediately. — Monitor air quality constantly with a burning candle.

the radio)	If oxygen becomes scarce (candle goes out): ventilate thoroughly for a short time and close air vents again. — Await further information.
Radiation warning i.e., start of radioactive fall-out (personal perception, information over the radio or local alert)	— Keep ventilation in operation sufficiently to maintain adequate air quality (avoid extreme air movements in the makeshift shelter). — Close all other openings (entrance, emergency exit). Check coverings of emergency exit/air vent (section 3.3, Fig. 12). — Do not leave the makeshift shelter or underground rooms adjoining it. — Await further information.
Atomic attack warning i.e., immediate use of atomic weapons expected (information over the radio, local alert)	— Check measures taken to protect against blast and collapse (props, bracing). — Close doors and openings. — Maintain ventilation of shelter. — Do not leave shelter, lie or sit on the floor as far as possible, do not lean on walls or props/supports.
Attack with conventional weapons NB – Chemical warfare agents could be used in any attack. See 'Chemical attack warning'. (Personal perception of explosions in the vicinity, information over the radio, local alert)	— Stop any artificial ventilation, ventilate makeshift shelter naturally. — Close all openings if air becomes very dusty (check air quality with a burning candle). — Check measures taken to protect against blast/collapse and fragments (props, bracing, openings). — Remain in the makeshift shelter, lying or sitting on the floor as far as possible, do not lean against walls or supports.
Entrance blocked by rubble/debris	— Leave the makeshift shelter by escaping through the emergency exit or clear a path (see section 4.5 for tools to facilitate escape). — Do not leave the makeshift shelter unless life is endangered by staying in it, or danger outside has passed.
Fire outside the makeshift shelter in the building above it or in the immediate vicinity. (Personal perception – smell of burning/smoke)	— If hot air or smoke comes in through the air vents, close all vents/openings (check air quality with candle). — In the event of a shortage of oxygen, leave the makeshift shelter, heading for the adjacent rooms or the outside. NB – The gas mask used to protect against chemical warfare agents *will not protect* against the gases produced by a fire (CO).
Fire inside the makeshift shelter	— Fight the fire immediately it is discovered using the means provided (section 4.5). — If necessary, stay in nearby rooms/areas until the makeshift shelter is 'habitable' again.

7. NOTES ON POSSIBLE MEANS OF PROTECTION WHEN UNDERGROUND FACILITIES ARE NOT AVAILABLE

In buildings with basements, the latter should be used primarily as makeshift shelters. This applies even if some people have to seek shelter outside their own dwellings, in neighbouring buildings or even further afield. In most cases, underground premises are preferable to buildings above ground. Where such underground premises are non-existent, considerable protection can sometimes be found if the place of refuge in buildings above ground is chosen correctly.

When selecting premises above ground to provide makeshift shelter, consideration must be given to the weapons effects against which such a place of refuge can provide protection, namely:

— the heat emitted by nuclear weapons,
— flying fragments, in particular splinters of glass, projected at high speed,
— radioactive fall-out.

It is relatively easy to understand the dangers of the first two categories, but it is difficult to visualize the danger from radioactive fall-out. Radioactive fall-out should be likened to fine sand (not dust!) with the same grain size as normal caster sugar, which falls from the sky following a nuclear explosion in the vicinity of the ground. It settles on roofs, balconies, trees and on the ground and emits radiation from where it lands with declining intensity.

So makeshift shelters in parts of buildings above ground are primarily threatened by the radiation emitted downwards from the roof and by the radiation from the ground (or from the roofs of neighbouring buildings) produced by the radioactive 'sand' lying there. Figure 44 below shows the effect of the radiation and appropriate protective measures in diagrammatic form.

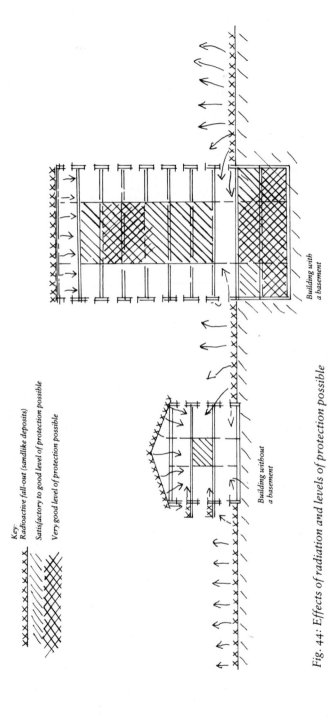

Key:
Radioactive fall-out (sandlike deposits)

Satisfactory to good level of protection possible

Very good level of protection possible

Building without
a basement

Building with
a basement

Fig. 44: Effects of radiation and levels of protection possible

From this it is clear that a makeshift shelter in parts of buildings above ground must satisfy the following requirements:

— no direct visual communication with the outside;
— walls and ceilings/floors must be as solid as possible and the makeshift shelter must be as far as possible from the world outside;
— there must be as little combustible material as possible in the vicinity of the makeshift shelter (if possible, building should be of fireproof construction or protected against fire).

When applied in practice, these rules mean that protection can be secured in buildings above ground in the following places (the best places are indicated first, followed by those offering a lesser amount of protection):

1. Heart (core zone) of fireproof buildings (even the upper stories of the core zone are suitable except for the areas immediately underneath the roof).
2. Heart of conventional buildings. The place of refuge should be as far away from the roof and the surrounding ground surface (or the roofs of neighbouring buildings) as possible.
3. Heart of light dwelling structures.

Protection can be improved considerably by improvising additional screening and other measures. In this respect, we would refer you to the relevant specialist literature, mentioning the corresponding US publications in particular (e.g., *In Time of Emergency – H–14*, reprinted 1980, contact: Office of Public Affairs, Federal Emergency Management Agency, Washington D.C. 20472).

APPENDIX 1

Example of Assessment for a Large Makeshift Shelter in an Underground Car Park

Planning Documents
1. System for assessment of suitability for use as a makeshift shelter
2. Sketch giving location and data on surroundings
3. Sectional/ground plan showing reinforcements
4. Description of conversion work and calculation of material requirements
5. List of materials required
6. List of times required

1.1 *Assessment of suitability for use as a makeshift shelter*

for the   *Seldwyla*   locality/sector.
Planning year: *1976* Carried out at the time of: *Makeshift shelter planning*

Information about Premises Surveyed
Street/location/owner: *Bahnhofstrasse 103, Property management company*
Type of building: *Apartment and office block, fireproof reinforced concrete structure*
Peacetime use of makeshift shelter premises: *Garage/car park*
Floor area/volume of makeshift shelter premises: *220* sq.m   *517* cu.m
Suitability of premises for use as
makeshift shelter (cf. sheet 3) : GOOD   Capacity: *110* places

---

(Cross appropriate boxes, 1 cross for each numbered section)

Assessment Criteria	Rating			
	v. good	good	usable	poor
Risks (sections 1–2)				
(1) *Fire risk constituted by building directly above*				
MAJOR FIRE RISK			☐	
(Quantities of combustible materials in the vicinity of the premises, i.e., — more than one wooden floor above the premises — building above of timber construction — combustible materials such as timber, furniture, plastics, hay, fuels above or near shelter)				
SLIGHT FIRE RISK	☒			
(all cases not classified under Major Fire Risk, in particular: — single-family houses, excluding timber houses — residential and business premises of fire-proof construction not containing combustible materials in any quantity)				
NO BUILDING ABOVE THE PREMISES	☐			
(2) *Danger from the surroundings*				
MAJOR RISK FROM FIRE AND RUBBLE				☐
(Means of escape non-existent or inadequate, e.g., because of building density, gap between buildings less than building height. Example: Old parts of cities, old village centres)				

Assessment Criteria	Rating			
	v. good	good	usable	poor
MODERATE RISK FROM FIRE AND RUBBLE (Escape more difficult due to surrounding buildings mainly constituting major fire risk. Gaps between buildings more than building height. Examples: older apartment blocks, industrial areas)			☐	
SLIGHT RISK FROM FIRE AND RUBBLE (Good means of escape, surrounding buildings mainly constitute slight fire risk. Examples: areas with mainly one-family houses, modern housing estates) **NO RISK**	☐	☒		
Construction of premises (sections 3–6)				
(3) *Ceiling*				
ROCK	☐			
REINFORCED CONCRETE CEILING (min. 14 cm thick)	☒			
VAULTED CEILING (brick)		☐		
TIMBER BEAMED CEILING				☐
OTHER (Specify:)			☐	☐
(4) *Surrounding walls* (mainly)				
ROCK	☐			
REINFORCED CONCRETE WALLS	☒			
CONCRETE WALLS		☐		
BRICK WALLS			☐	
TIMBER CONSTRUCTION				☐
(5) *Floor*				
ROCK	☐			
CONCRETE FLOOR	☒			
THIN CEMENT FLOOR		☐		
NATURAL EARTH FLOOR			☐	
(6) *Partition walls* (mainly)				
NONE	☐			
CONCRETE OR REINFORCED CONCRETE	☒			
LIGHT PARTITION WALLS (timber, etc.)		☐		
BRICK PARTITION WALLS			☐	
Location of premises (sections 7–10)				
(7) *Proportion of premises underground*				
100%	☐			
approx. 80–100%		☒		
approx. 50–80%			☐	
less than 50%				☐

Assessment Criteria	Rating			
	v. good	good	usable	poor
(8) Exposure of shelter shell **(8.1) Building basements** Openings in ground floor or walls of directly adjoining rooms/areas: – NO openings in ground floor – Openings constitute LESS THAN 50% of the corresponding wall areas – Openings constitute MORE THAN 50% of the corresponding wall areas	☐	☒	☐	
(8.2) Free-standing premises ENTIRE ceiling area covered with earth (or made of concrete more than 30 cm thick)	☐			
Ceilings (less than 30 cm thick) MORE THAN 50% covered with earth		☐		
Ceilings (less than 30 cm thick) LESS THAN 50% covered with earth			☐	
(9) Openings in shelter shell Total opening area: Less than 6 sq.m 6–12 sq.m More than 12 sq.m	☐	☒	☐	
(10) Groundwater flooding in the event of a power failure? (i.e., due to a pumping system failing) No Yes	☒ ☐			
Number of crosses n (Check: must be 10 crosses) Weighting q	6 1	4 2	0 5	
$q \cdot n$	$6 \cdot 1 = 6$	$4 \cdot 2 = 8$	$0 \cdot 5 = 0$	
Number of points = total $q \cdot n$ ($q \cdot n$)	14			
Suitability of the premises				

Suitability rating	Number of points $q \cdot n$	Work involved
Very good	10–11	Little *
Good	12–20	Moderate amount
Usable	21–50	A lot
Poor (Such premises should only be used for makeshift shelters when nothing else is available)	over 50	

* Premises can be made ready for use in a short time.

1.2 Sketch giving location and data on surroundings

Draw at 1:500

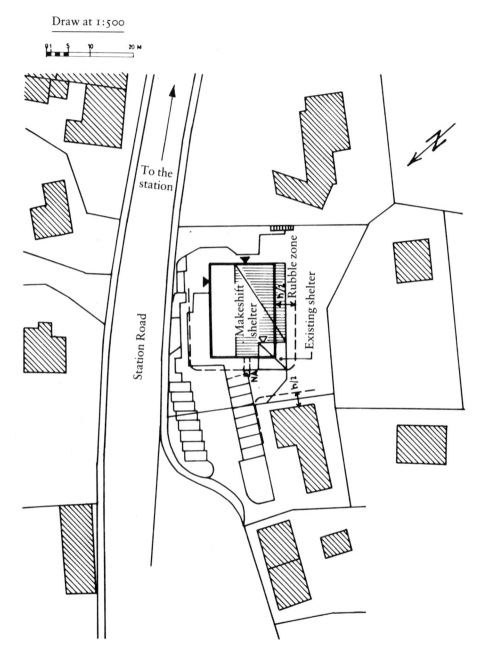

To the station

Station Road

Makeshift shelter

Rubble zone

Existing shelter

Fig. 45.

1.3 Sectional/ground plan showing reinforcement

Draw at 1:100

(1:50 if possible)

Key to Work Necessary

	Item	Work:
Adaptation phase 1	①	Seal up existing garage entrance and one window opening
	②	Seal up door openings
	③	Reinforce the entrance
	④	Prepare the emergency exit
	⑤	Prepare the air vents
Adaptation phase 2	⑥	Reinforce the ceilings
	⑦	Extend the emergency exit
	⑧	Improve the air circulation

Key to Ground Plan and Section

1	Ground plan
2	Makeshift shelter (F=220 sq.m, V=517 cu.m = 129 places)
3	Ventilated shelter – purpose-built (F=33 sq.m, V=77·5 cu.m = 31 places)
4	Reinforced concrete
5	Existing ventilation unit
6	Parking area
7	Covered
8	Shaft
9	San. distribution unit
10	Trough
11	Hall
12	Heating system
13	Electrical distribution panel
14	Archives
15	Cycle store
16	Washroom
17	Tank room
18	Sectional view

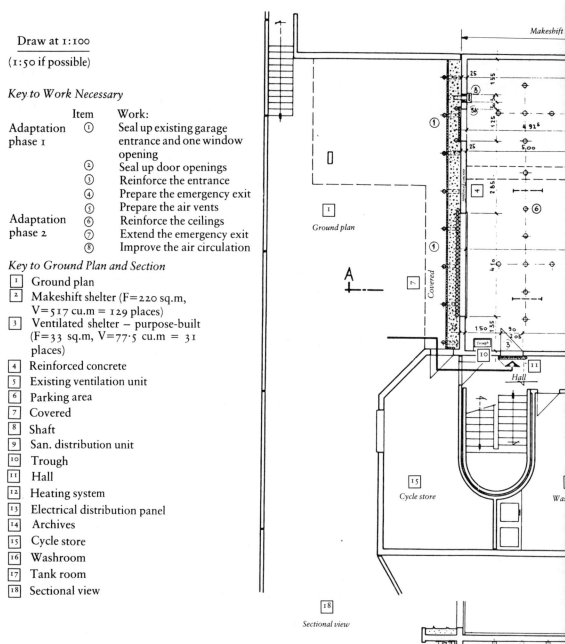

Ground plan

Sectional view

Fig. 46.

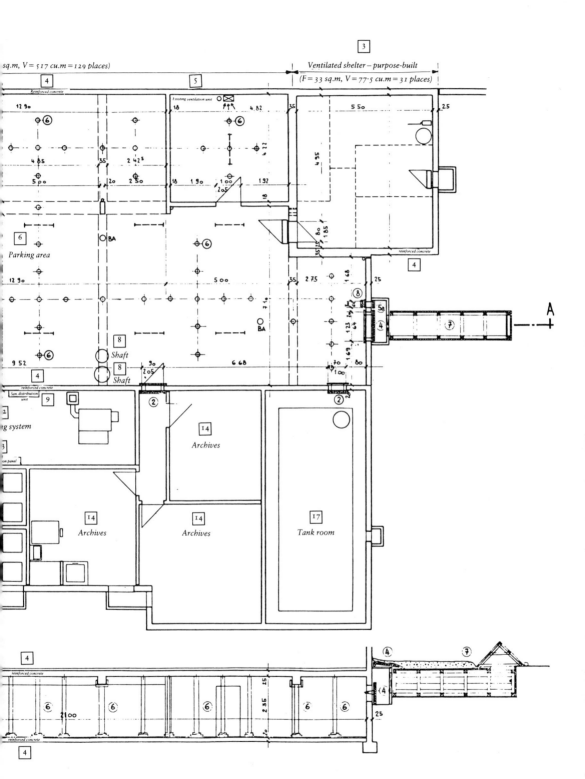

sq.m, V = 517 cu.m = 129 places)

⟨3⟩

Ventilated shelter – purpose-built

(F = 33 sq.m, V = 77·5 cu.m = 31 places)

⟨4⟩   ⟨5⟩

Reinforced concrete

12 90   *Existing ventilation unit*   18   4.82   35   5 50   25

4 85   35   2 42⁵

5 00   20   2 50   18   1 90   1 92   4 35

⟨6⟩   BA

Parking area   ⟨6⟩

12 90   5 00   35   2 75   168   25

⟨8⟩

⟨8⟩ Shaft   5a   A

9 52   ⟨6⟩   BA   ⟨4⟩ ⟨7⟩

⟨4⟩   ⟨8⟩ Shaft   90   6 68   70   80
                     2 05   1 00

reinforced concrete

San. distribution unit   ⟨9⟩   ⟨2⟩   ⟨2⟩

g system   ⟨14⟩   *Archives*

on panel   ⟨17⟩

⟨14⟩ *Archives*   ⟨14⟩ *Archives*   *Tank room*

⟨4⟩

reinforced concrete   15

⟨6⟩   2 35   ⟨4⟩

⟨6⟩   21 00   ⟨6⟩   ⟨6⟩   ⟨6⟩   ⟨6⟩   25

reinforced concrete

⟨4⟩

1.4 Description of conversion work and calculation of material requirements

A. Reinforcing work for adaptation phase 1

Item 1 – Sealing up existing garage entrance and one window opening and wall-reinforcement in area of garage exit

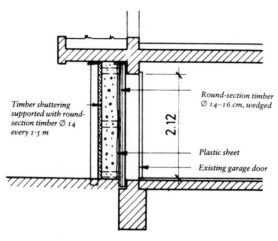

Timber shuttering supported with round-section timber Ø 14 every 1·5 m

Round-section timber Ø 14–16 cm, wedged

2.12

Plastic sheet

Existing garage door

Sectional view

Fig. 47: Sealing of garage door

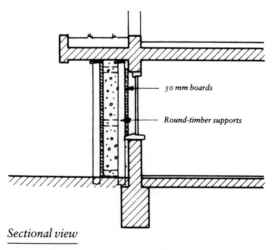

50 mm boards

Round-timber supports

Sectional view

Fig. 48: Sealing of window opening

Materials required:

— Boards $d=50$ mm 1.50×1.50
                                   $2.50 \times 11.50 = 31$ sq.m

— Round-section timber
    $\varnothing$ 14 cm 30×2.50
                                    $10 \times 2.50 = 100$ m

— Plastic sheet $5 \times 2.50 = 12.5$ sq.m
— Wooden wedges   approx. 80
— Earth/filling material $11.50 \times 0.50 \times 2.50 = 14$ cu.m

Item 2 – Sealing up door openings inside the building

1 opening to archives, size 0·90/2·05 m
1 opening to oil tank room, size 0·70/1·00 m

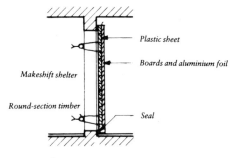

Sectional view

Fig. 49: Sealing of door opening

Apply nailed boards from exterior, 2 layers nailed together cross-wise with plastic sheet in between, with sealing strip on support. Attach with wire to round-section timbers (sealing up completely with solid material is not needed as the adjoining areas are not exposed).

Materials required:

— Boards $d=50$ mm $1.10 \times 2.10 \times 2$
                              $0.90 \times 1.20 \times 2 = 7$ sq.m

— Round-timber $\varnothing$ 10 cm 2×1.10
                                  $2 \times 0.90 \quad = 4$ m

— Plastic sheet   approx.   4 sq.m
— Sealing strip, adhesive tape   11 m
— Nails, wire

Item 3 – Reinforcing the entrance (internal entrance)

Fit an outward opening, reinforced door. Built as indicated in section 3.22, Fig. 10.
Size of opening 0·90/2·05 m

Materials required:

— Boards $d=50$ mm	$2 \times 1·10 \times 2·15 \cong 5$ sq.m
— Iron bar $\varnothing$ 50 mm incl. hinge plates, $1=2·50$ m	1
— Door securing means	1
— Plastic sheet/aluminium foil	2·5 sq.m
— Sealing strip	6 m
— Nails	

Item 4 – Preparing the emergency exit

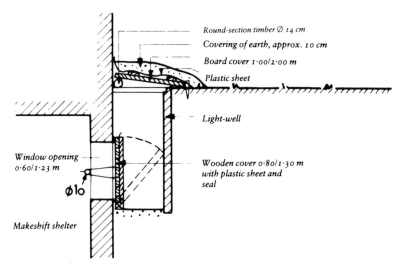

Round-section timber Ø 14 cm

Covering of earth, approx. 10 cm

Board cover 1·00/2·00 m

Plastic sheet

Light-well

Window opening 0·60/1·23 m

ø10

Wooden cover 0·80/1·30 m with plastic sheet and seal

Makeshift shelter

Sectional view

Fig. 50: Emergency exit

Materials required:

— Boards *d* 50 mm	2×0·80×1·30
	1·50×2·00=5 sq.m
— Round-section timber Ø 10 cm	4 m
Ø 14 cm	3 m
— Plastic sheet	1·00×1·50
	1·50×2·50≅5 sq.m
— Sealing strip	~5 m
— Earth/filling material	0·2 cu.m
— Nails	

Item 5 – Air vents (closable)

5a – Wooden cover to close existing 0·30×0·60 m opening for
emergency exit

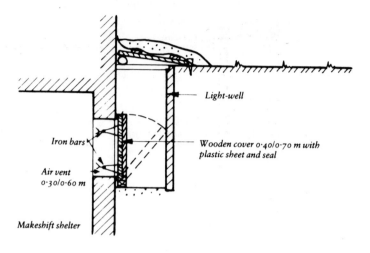

Sectional view

Fig. 51: Air vent for emergency exit

Materials required:

— Boards $d = 50$ mm $2 \times 0·40 \times 0·70 = 0·6$ sq.m
— Plastic sheet   approx.   0·5 sq.m
— Iron bar $\varnothing$ 10 to 14 mm   1·5 m
— Sealing strips   approx.   3·0 m
— Nails

5b – Ventilation pipe to provide air vent when wall is reinforced

Lay a pipe approx. 150 to 200 mm in diameter in the existing air vent and run it through the wall reinforcement. Includes sealing plug to close off air vent hermetically.

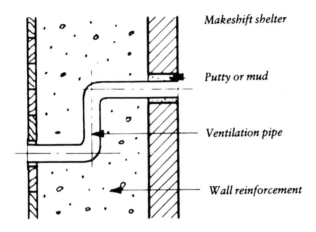

Makeshift shelter

Putty or mud

Ventilation pipe

Wall reinforcement

Sectional view

Fig. 52: Air vent for emergency exit

Materials required:

— Pipe Ø 150–200 mm with two 90° bends   approx. 1·30 m
— Putty or loam (mud)
— Sealing plug

B. *Reinforcing work for adaptation phase 2*

Item 6 – Ceiling reinforcement

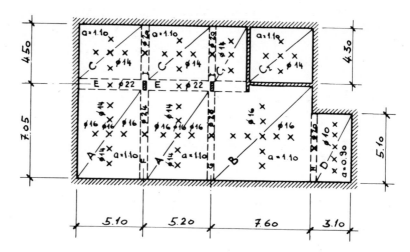

Ground plan

Fig. 53a: Ceiling reinforcement

Ceiling system: reinforced concrete ceiling with beams and supporting columns. Existing ceiling thickness $d=25$ cm.

Establishing cross-sections and arrangement of additional props:

Span A: $b:1=5\cdot10:7\cdot05=1:1\cdot38>1:1\cdot5$
→ extra props in both directions
$a\cong1\cdot10$ m, $\emptyset$ 14 to 16 cm.

Span B: $b:1=7\cdot05:7\cdot60=1:1\cdot07>1:1\cdot5$
→ extra props in both directions
$a\cong1\cdot10$ m, $\emptyset$ 16 cm.

Span C: $b:1=4\cdot50:5\cdot20=1:1\cdot16>1:1\cdot5$
→ extra props in both directions
$a\cong1\cdot10$ m, $\emptyset$ 14 cm.

Span D; $b:1=3\cdot10:5\cdot10=1:1\cdot65<1:1\cdot5$
→ extra props in one direction only
$a\cong0\cdot90$ m, $\emptyset$ 10 cm.

Beam E: $P=\dfrac{5\cdot20}{2}\cdot\dfrac{5\cdot75}{2}\cdot5\cdot7=42\cdot6\,\text{t}\rightarrow\emptyset$ 22 cm

Beam F: $P= \dfrac{5\cdot20}{2}\cdot\dfrac{7\cdot0}{2}\cdot5\cdot7=51\cdot9\text{ t}\rightarrow\varnothing\ 24\text{ cm}$

Beam G: $P= \dfrac{6\cdot50}{2}\cdot\dfrac{7\cdot0}{2}\cdot5\cdot7=64\cdot8\text{ t}\rightarrow\varnothing\ 26\text{ cm}$

Beam H: $P= \dfrac{5\cdot30}{2}\cdot\dfrac{5\cdot10}{2}\cdot5\cdot7=38\cdot5\text{ t}\rightarrow\varnothing\ 20\text{ cm}$

Beam I: $P= \dfrac{5\cdot20}{2}\cdot\dfrac{4\cdot50}{2}\cdot5\cdot7=33\cdot3\text{ t}\rightarrow\varnothing\ 20\text{ cm}$

No additional foundations are needed under the props as the floor slab is more than 15 cm thick.

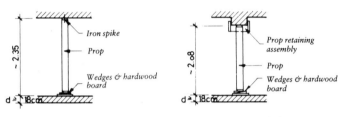

Prop assembly in ceiling slab spans · Prop assembly under the beams

Fig. 53b: Sectional view of ceiling reinforcement

Materials required:

— Round-section
   timber

$\varnothing$ 10 cm	$5\times2\cdot30=$	12 m
$\varnothing$ 14–16 cm	$41\times2\cdot30=$	95 m
$\varnothing$ 20 cm	$3\times2\cdot00=$	6 m
$\varnothing$ 22–24 cm	$3\times2\cdot00=$	6 m
$\varnothing$ 26 cm	$1\times2\cdot00=$	2 m

— Hardwood wedges   approx.   140 off

— Hardwood boards

$30\times30\times3$ cm =	48 off
$40\times40\times3$ cm =	5 off

— Boards $d=30$ mm (to secure
   props)   approx.   5 sq.m

— Round iron rod $\varnothing$ 10–14 mm,
   $1=20$ cm   approx.   50 off

— Nails

Item 7 – Extending the emergency exit
Prepare a shored trench with covering, 1×5 m, as escape
tunnel.

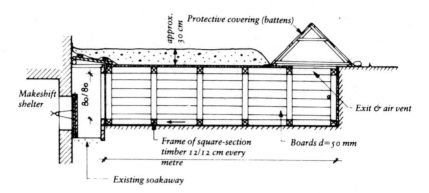

Sectional view

Fig. 54: Emergency exit

Materials required:

— Boards *d*=50 mm

$$4{\cdot}0\times1{\cdot}5$$
$$2\times5{\cdot}0\times1{\cdot}0$$
$$1{\cdot}0\times1{\cdot}0\cong17 \text{ sq.m}$$

— Angle-section timber   12×12 cm   25 m

— Battens   approx. 20 m

— Plastic sheet   5 sq.m

— Nails

— Concrete – opening made
in light-well $0{\cdot}80\times0{\cdot}80\times0{\cdot}1=0{\cdot}07$ cu.m

— Excavation $1{\cdot}0\times1{\cdot}0\times5{\cdot}0=5$ cu.m

— Earth covering $1{\cdot}5\times0{\cdot}3\times4\sim2$ cu.m

Item 8 – Improving the air circulation (installation of fans)
Total needed: 2
Extractor units with integral fans from kitchens on 2nd floor.

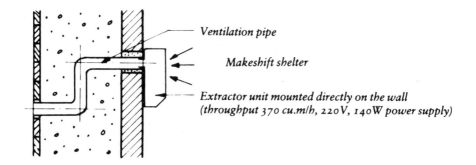

Ventilation pipe

Makeshift shelter

Extractor unit mounted directly on the wall
(throughput 370 cu.m/h, 220V, 140W power supply)

Sectional view

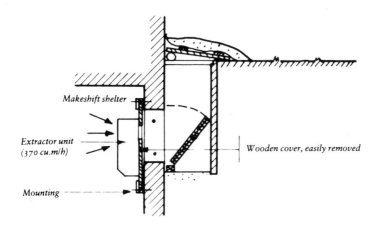

Makeshift shelter

Extractor unit
(370 cu.m/h)

Wooden cover, easily removed

Mounting

Sectional view

Fig. 55: Fan installation

75

Materials required:

— Existing extractor units	(370 cu.m/h)	2 off
— Wooden board $d=30$ mm	0·40×0·70 m	1 off
— Battens		2·5 m

Further remarks:

There is no need to spread earth over the floor immediately above the makeshift shelter as the existing ceiling/floor structure is sufficient to provide the desired amount of protection (25 cm of reinforced concrete with a 6–10 cm structure on top).

1.5 List of materials required

Material		Total	For adaptation phase 1	PLUS for adaptation phase 2
Timber materials				
— Boards	$d = 50$ mm	66·0 sq.m	49·0 sq.m	17·0 sq.m
	$d = 30$ mm	6·0 sq.m	–	6·0 sq.m
— Round-section timber[1]	Ø 10 cm	20·0 m	8·0 m	12·0 m
	Ø 14–16 cm	198·0 m	103·0 m	95·0 m
	Ø 18–20 cm	6·0 m	–	6·0 m
	Ø 22–24 cm	6·0 m	–	6·0 m
	Ø 26 cm	2·0 m	–	2·0 m
— Angle-section timber	12/12 cm	25·0 m	–	25·0 m
— Hardwood boards	30/30/3 cm	48 off	–	48 off
		5 off	–	5 off
— Hardwood wedges		220 off	80 off	140 off
— Battens		23·0 m	–	23·0 m
Miscellaneous, ironmongery				
— Plastic sheeting		30·0 sq.m	25·0 sq.m	5·0 sq.m
— Sealing strip		24·0 m	24·0 m	–
— Round iron bar	Ø 10–14 mm	50 off	–	50 off
— Ventilation pipe	Ø 150–200 mm	1·30 m	1·30 m (incl. 2×90° bends)	
— Iron bar (pipe?) including 3 flat iron strips and 3 pipe clamps	Ø 50–100 mm	2·50 m	2.50 m, 1 off	–
— Wire		20·0 m	10·0 m	10·0 m
— Nails	$l = 100$–140 mm	20 kg	15 kg	5 kg
— Putty, loam (mud)		10 kg	10 kg	–
— Dogs		10 off	–	10 off
Earth/filling material		16·0 cu.m	14·0 cu.m	2·0 cu.m

[1] Prepare separate itemized list

1.6 *List of times required*

	Item in conversion schedule	Reinforcing/conversion work	Estimated time needed Mh
Adaptation phase 0	–	Clear makeshift shelter	15
Adaptation phase 1	– 1 2 3 4 5 –	Preparatory work for adaptation phase 1 (preparation of materials excluding actual procurement of materials, etc.) Sealing up garage entrance, incl. 1 window opening and wall reinforcement Sealing up door openings Reinforcing the entrance Preparing emergency exit Air vents Closing windows, shutters, blinds and doors in adjoining and ground floor rooms/areas (organizational measures)	30 120 6 10 15 3 2
		Time needed for adaptation phases 0+1	200 Mh
		per sq.m of shelter floor area $(\frac{200}{220})=$	0·9 Mh/sq.m
Adaptation phase 2	– 6 7 8	Preparatory work for adaptation phase 2 Propping the ceiling including securing of props (50 props) Extending the emergency exit (all calculations assume manual labour) Improving the air circulation	 75 50 5
		Extra time needed for adaptation phase 2	150 Mh
		Total time needed for adaptation phases 1+2 per sq.m of shelter floor area	350 Mh 1·6 Mh/sq.m

APPENDIX 2

Weapons Effects and Protection Afforded
The effects of the most important modern weapons and the protection that can be provided by a well-constructed makeshift shelter are described below in a highly simplified manner.

2.1 ATOMIC WEAPONS

The explosion is accompanied by the emission of extremely intense heat and light, lasting a second or so. The primary nuclear radiation begins immediately. The pressure wave reaches the shelter a few seconds after the explosion. This produces a wind many times the force of a hurricane. This lasts as long as the overpressure acting all around, i.e., a few tenths of a second, or a few seconds in the case of relatively large devices. At the same time, large quantities of debris are hurled through the air. The heat given off can ignite inflammable materials. Following an explosion near the surface, the radioactive fall-out starts after a quarter of an hour or so, and its effects can last for days or weeks. Now what are the specific weapons effects and what protection can be provided?

2.11 *Mechanical weapons effects*
The mechanical weapons effect essentially consists of the blast. This is a factor of the distance of the shelter from the centre of the burst, the size of the device and the height at which the burst occurs. At one particular point the blast produces a very sharp rise in air pressure up to a peak, followed by an initially steep, then gradual decline which lasts a few tenths of a second. This overpressure phase is followed by a longer phase of relatively slight underpressure.

The graph below shows the approximate distances at which different sizes of device produce a peak pressure of 1 bar and 0·5 bar overpressure by way of example when exploded near the surface. For comparison: The devices exploded in 1945 were of 12 KT over Hiroshima and 22 KT over Nagasaki. (A good makeshift shelter can withstand an overpressure of roughly 0·5 bar.)

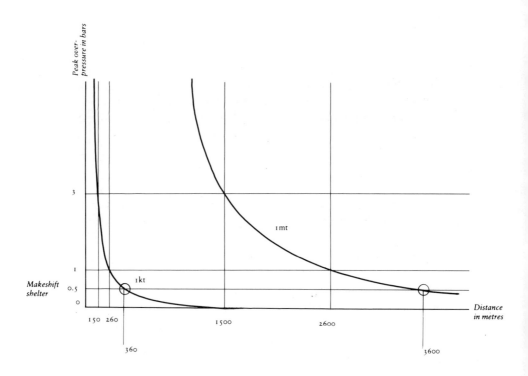

Fig. 56: Maximum pressure as a function of the distance from a surface burst

The blast is 'reflected' by structures above ground and by exposed parts of the shelter shell. This increases the maximum strength of the pressure wave two or more times over. In the ground the blast produces a kind of earthquake, the so-called air-induced earth tremor. This air-induced earth tremor produces the pressure loadings on the side walls and shakes the entire shelter. Here, the intensity of the shaking depends a great deal on the nature of the ground but is also determined by the shelter itself.

Optimum protection against the mechanical effects of nuclear weapons can be obtained if

— as far as possible, all the makeshift shelter lies underground,
— fittings and equipment are limited to the essentials required for survival.

2.12 Primary nuclear radiation

The primary nuclear radiation spreads – invisible to the human eye – from the rapidly expanding fireball of the explosion. This radiation is the sum of all the directly and indirectly ionizing rays emitted by the fireball and the atomic cloud (mushroom) within a minute. Only two components of this radiation pose a significant threat to the shelter and its occupants: the gamma rays and the neutron rays.

Primary nuclear radiation is propagated in a similar fashion to light. Depending on whether the rapidly climbing fireball is in the line of vision from the shelter or not, the primary radiation impinges directly or indirectly on the shell of the shelter. The intensity of the primary nuclear radiation is described by the so-called dose, i.e., the amount of the radiation absorbed by a damaged medium. In physical terms, this means the radiation energy retained by the absorbing medium per unit of mass, measured in rads (1 rad = 100 erg/g).

A so-called equivalent dose, measured in rem, is used to define the damaging effect of the radiation in the tissue of the human body. This equivalent dose takes account of the different biological effects of the gamma and neutron rays and of other factors, allowing, among other things, for the depth of penetration into the body.

At the distance from a nuclear explosion at which a blast of 1 bar or 0·5 bar is produced, the radiation dose depends a lot on the size of the device. At this range we can assume that an unprotected human being would receive a total equivalent dose D_o of about 10,000 rem and about 3000 to 4000 rem respectively. This is far above the lethal dose.

Protection against the effects of the primary nuclear radiation is provided by the underground location, the earth embankments and the earth covering or the building above the shelter, the structure of the shelter shell (concrete) and expedient arrangement of the entrances. In many cases, the result can be to reduce the effect of the radiation to the practically harmless dose of at most 100 rem.

2.13 Secondary nuclear radiation (fall-out)

Secondary nuclear radiation can almost only occur after explosions near the surface. It emanates from small particles of dust containing highly radioactive fission products of the nuclear explosion. Carried by the wind, this dust can contaminate a large area. Radioactive contamination can be spread over hundreds of kilometres. Radioactive fall-out is visible. The intensity of the radiation emitted declines as a factor of the distance from the burst and the passage of time after the burst.

Secondary radiation can be reduced to fractions of its initial level above all by solid screening materials such as concrete, earth, etc. Good makeshift shelters, even ordinary enclosed underground rooms (windowless cellars), guarantee survival through radioactive fall-out. However, in this case, one must expect to remain in the shelter for days or weeks. Nobody should leave the shelter until the radiation level has dropped sufficiently.

2.14 *What is the neutron bomb?*

The neutron bomb is a small nuclear weapon used specifically against formations of armoured vehicles. It acts mainly through the initial radiation (neutron radiation accompanied by gamma radiation) produced during nuclear fusion. Because of its design, the blast and heat generated by this bomb are reduced considerably and hardly any radioactive fall-out is produced. The initial radiation penetrates almost all known vehicle armour systems.

Good makeshift shelters located underground provide far better protection than the armour used for fighting vehicles because of their construction (concrete), the earth covering and earth embankments and the building above. Building materials such as concrete and earth are particularly effective against neutron radiation.

2.15 *Thermal radiation*

Around a third of the energy of a nuclear explosion is released in the form of heat. This can ignite combustible materials at great distances and cause fires in buildings in particular. This thermal radiation is one of the greatest dangers as far as unprotected people are concerned. For people in makeshift shelters wholly or mostly underground, the thermal radiation is of hardly any significance. However, secondary dangers affecting the shelter may arise through a building standing directly above it or in the immediate vicinity being set alight.

2.2 CONVENTIONAL WEAPONS

(i.e., devices other than nuclear, chemical or biological weapons)

2.21 *Explosive projectiles*

Conventional weapons which might pose a threat to shelters are primarily artillery shells/projectiles and airborne bombs containing conventional explosives. The effect of these weapons is characterized firstly by penetrating capacity and secondly by blast and fragmentation when their charge explodes.

Projectiles or bombs that detonate on impact and in particular those with proximity detonators are effective *above* the ground almost entirely. They do not pose a real threat to shelters of conventional design or to makeshift shelters. Delayed-action fuses are normally used to make projectiles and bombs explode inside buildings, causing considerable destruction. Weapons with delayed-action fuses could penetrate most shelters.

It is often argued that attacks such as those on German cities during the Second World War could pose a threat to shelters because of the saturation bombing methods used. However, during the Second World War the primary objective of these raids was to set the cities alight, which was often achieved with the old city buildings standing at that time. The high-explosive bombs were used primarily to 'loosen up' the buildings, the incendiaries to start innumerable fires. In any future war, such bombardments using carpet bombing methods are highly unlikely to occur because

— the required number of bombers will not be available and fighter-bombers will be used for other targets,
— the bombers would be extremely vulnerable to anti-aircraft weapons,
— modern cities are much less vulnerable to such attacks, architecturally speaking,
— nuclear weapons offer cheaper, quicker and more reliable means for calculated destruction of towns and cities.

Flat-trajectory weapons (e.g., projectiles fired by tanks) do not pose any real threat to makeshift shelters given that they lie underground.

Chemical warfare agents are substances which, in minute quantities, have an irritating or toxic effect on humans, animals and plants. Chemical warfare agents do not damage buildings, materials, installations and the like. Chemical warfare agents are mainly used in the course of a conventional attack on specific military targets as bases and command centres. They can be used in two ways:

For so-called 'non-persistent' applications, aerosol or gaseous agents are used primarily; this method is generally employed for surprise attack. The effect only lasts for a short time (a few hours at most); the area is not contaminated. The immediate area in which the agents are used extends to 1 sq.km at most; however, the cloud of chemical agent that forms can be carried by the wind and may pose a threat to areas of as much as 100 sq.km.

For so-called 'persistent' applications, liquid or solid warfare agents are mainly employed over an area of up to about 1 sq.km; the effect lasts a relatively long time (days to weeks); the area is contaminated but the chemical warfare agent only spreads to areas outside the area in which it has been used to a limited extent.

In 'non-persistent' applications, humans are mainly poisoned by inhaling chemical warfare agents, in 'persistent' applications primarily through contact with contaminated objects and through the evaporation of the toxic liquids. As far as civil defence is concerned, the most important chemical warfare agents are the nerve poisons which can be employed 'non-persistently' or 'persistently'. Nerve poisons kill humans and animals within a few minutes. Depending on the concentration, just a few breaths are lethal to unprotected human beings.

Protection against chemical warfare agents is obtained by ensuring that the occupants of shelters do not come into contact with these dangerous substances and, above all, do not inhale them. In the case of makeshift shelters this is achieved by making the shell as airtight as possible and, above all, by taking the personal precaution of wearing a protective gas mask.

2.4 BIOLOGICAL WEAPONS

By biological weapons we mean the use of germs (bacteria and viruses) harmful to human, animal and even plant life. Little is known about biological warfare. Deliberate biological attacks are unlikely since they would have incalculable consequences and side-effects for an aggressor as well. Moreover, problems arise with regard to their use (because of the question of the viability of the germs) and their dispersion. Biological weapons can be made which cause cholera, dysentery, plague, typhus, anthrax, flu. In many cases, the disease breaks out days after infection.

These weapons could be used like chemical warfare agents or introduced into water supplies, foodstuffs, etc., by saboteurs. The problem is that the aggressor can inoculate specifically against the diseases concerned before the attack.

A makeshift shelter *per se* does not give any protection against biological weapons. The primary means of defence is to make sure that no germs can be brought into shelters through foodstuffs, water, contamination, etc., if occupants have to leave the shelter. These dangers can be limited by strict hygiene and minimal contact with occupants of other shelters. Inoculation is only possible if the germs to be used in a probable attack are known in advance.

2.5 SECONDARY WEAPONS EFFECTS

2.51 *Rubble and debris*

When nuclear weapons are used or attacks with conventional weapons, buildings and their installations are totally or partially destroyed. The fragments and debris produced in the process are hurled in all directions and pose the greatest threat to unprotected people. The spread and density of the debris and the sort of piles of rubble produced are largely determined by the nature and density of buildings above ground and the size of weapon employed.

The reinforced makeshift shelters can take the load of rubble and debris produced when buildings collapse. Consequently, the air vents and emergency exits should be located as far from buildings as possible and, in the case of relatively large shelters, disposed on more than one side of the shelter. It cannot be over-emphasized that nobody must leave the shelter immediately after an attack – in contrast to the procedure in many air raid shelters during the Second World War. Even a badly damaged shelter can ensure survival for a long time. Plenty of time is available for escape. Fresh air can get to the shelter through relatively large amounts of rubble and debris.

2.52 *Risk of fire*

The shelter may be threatened if a building standing directly above or in the immediate vicinity of the shelter is burning. Fires of this kind can heat up the shell of the shelter sufficiently to make the temperature inside the shelter unbearable. Apart from this, toxic fumes from the fire (CO, CO_2) may get into the shelter through openings.

Severe fire risks should be avoided by selecting the location for the makeshift shelter accordingly.

2.6 SUMMARY

2.61 *Physiological conditions for survival*

Research into the behaviour of man in extreme situations (in wars, crises, and tests) shows that it is possible to survive days or weeks in civil-defence shelters if the following *physiological conditions* are fulfilled in this order of priority:

86

1. Protection from weapon effects.
2. Protection from cold and wet weather.
3. Breathing air:
   A minimum of 18% (by volume) of oxygen and a maximum of 1% (by volume) of carbon dioxide are safety limits.
4. Tolerable shelter climate:
   i.e. about 29°C (84°F) at 100% relative humidity (for a healthy general population).
5. Sufficient drinking water:
   2 to 4 litres (0·52 to 1·05 gallons) (presupposing dry food), depending on the shelter climate; almost no washing water.
6. Adequate food supplies:
   8,700 kJ (2,100 kcal) of food per day prevent weight loss assuming minimum personal exercise.
7. Adequate sleeping accommodation:
   A sleeping space for each shelteree having an area of about 0·70 × 1·90 m (2·3 × 6·2 ft) and a clear height of 0·60 m (2 ft); with a noise level not exceeding 65 dBA and lighting (for reading) at 30 Lux plus a small emergency light.
8. Adequate space standards:
   A total of 1 m² (10·8 sq.ft) and 2·5 m³ (88·4 cu.ft) per shelteree.
9. Separate toilets:
   1 dry toilet per 30 persons.
10. Medication:
   For sleeping, headaches, colds, and gastrointestinal diseases; first-aid kit.

2.62   *Psychological behaviour*

The *psychological behaviour* of man in crises and war is characterized:
1. In peacetime
   — by repressing fear of a possible war (or catastrophe),
   — by grossly underestimating his ability to resist and survive,
   — by ignorance of and missing confidence in modern protection systems against modern weapon effects,
   — by taking insufficient precautions against wars and/or catastrophes.

2. In the pre-attack phase
   — first pre-attack phase
      — by a reluctance of leadership to order the commissioning and occupation of shelters,
      — by a reasonable behaviour of the population, provided there has been sufficient prior information on possible events,
      — by rather unreasonable behaviour (but not panic) in case of inadequate information,
      — by the readiness to occupy the shelters and stay there, if the population feels the danger and if the shelters are well built, equipped, and managed.
   — following pre-attack phases:
      — by a marked general readiness to take shelter, especially if the weapon effects were felt intensively in the first pre-attack phase,
      — by an increasing tendency of some individuals to fail to adhere strictly to civil defence rules, if the first attack was weaker than anticipated.
3. In the attack phase, the post-attack phase, and the recovery phase:
   — by the lack of panic in and outside the shelters, provided that survival is not directly and immediately threatened (e.g. by the imminent collapse of the shelter, by the penetration of dense smoke, or by unbearable heat in the shelter),
   — by adaptation to the situation, i.e. generally reasonable behaviour under stress combined with an endeavour to alleviate fear by boasting, frivolity, and cheerful conversation, if the attack did not approach the limit of survival,
   — by a great need to receive news and to inform others about one's own experiences.